Sustainable Civil Infrastructures

Sustainable Civil Infrastructures (SUCI) is a series of peer-reviewed books and proceedings based on the best studies on emerging research from all fields related to sustainable infrastructures and aiming at improving our well-being and day-to-day lives. The infrastructures we are building today will shape our lives tomorrow. The complex and diverse nature of the impacts due to weather extremes on transportation and civil infrastructures can be seen in our roadways, bridges, and buildings. Extreme summer temperatures, droughts, flash floods, and rising numbers of freeze-thaw cycles pose challenges for civil infrastructure and can endanger public safety. We constantly hear how civil infrastructures need constant attention, preservation, and upgrading. Such improvements and developments would obviously benefit from our desired book series that provide sustainable engineering materials and designs. The economic impact is huge and much research has been conducted worldwide. The future holds many opportunities, not only for researchers in a given country, but also for the worldwide field engineers who apply and implement these technologies. We believe that no approach can succeed if it does not unite the efforts of various engineering disciplines from all over the world under one umbrella to offer a beacon of modern solutions to the global infrastructure. Experts from the various engineering disciplines around the globe will participate in this series, including: Geotechnical, Geological, Geoscience, Petroleum, Structural, Transportation, Bridge, Infrastructure, Energy, Architectural, Chemical and Materials, and other related Engineering disciplines.

SUCI series is now indexed in SCOPUS and EI Compendex.

More information about this series at http://www.springer.com/series/15140

Yong Liu · Sabatino Cuomo ·
Junsheng Yang
Editors

Advances in Innovative Geotechnical Engineering

Proceedings of the 6th GeoChina International Conference on Civil & Transportation Infrastructures: From Engineering to Smart & Green Life Cycle Solutions – Nanchang, China, 2021

Springer

Editors
Yong Liu
School of Water Resources
and Hydropower Engineering
Wuhan University
Wuhan, China

Sabatino Cuomo
University of Salerno
salerno, Italy

Junsheng Yang
Central South University
Changsha, China

ISSN 2366-3405 ISSN 2366-3413 (electronic)
Sustainable Civil Infrastructures
ISBN 978-3-030-80315-5 ISBN 978-3-030-80316-2 (eBook)
https://doi.org/10.1007/978-3-030-80316-2

This Springer imprint is published by the registered company Springer Nature Switzerland AG
The registered company address is: Gewerbestrasse 11, 6330 Cham, Switzerland

Introduction

This volume contains 12 papers that were accepted and presented at the 6th GeoChina 2021 International Conference on Civil & Transportation Infrastructures: From Engineering to Smart & Green Life Cycle, held in NanChang, China, September 18 to 19, 2021. This volume presents the advances in innovation geotechnical engineering such as microscopic analysis, dynamics property, ground improvement method, application of machine learning in engineering, effect of temperature on soil characteristics and some other geotechnical issues which are quite relevant in today's world. It is anticipated that this volume can represent the advanced research direction on geotechnical engineering.

Contents

About the Editors

Prof. Yong Liu is currently a professor in the School of Water Resources and Hydropower Engineering and vice-director of the Institute of Engineering Risk and Disaster Prevention, Wuhan University. In the area of geotechnics, he has a strong interest in seepage problems in dam embankments, uncertainty quantification, as well as efficient stochastic analysis methods and applications of large-scale geotechnical structures. He published more than 90 academic papers in Géotechnique, ASCE-JGGE, and other peer-viewed journals, among which three were selected as ESI highly cited papers. In 2017, he won the Best Paper Award from the International Conference on Transportation Infrastructure and Materials. He has been serving as associate editor and editorial board member of several international journals.

Prof. Sabatino Cuomo Department of Civil Engineering, University of Salerno, Italy

Inception of debris avalanches: a Material Point Method modelling

- Coordinator of LARAM School (International School on "LAndslide Risk Assessment and Mitigation) for PhD students, http://www.laram.unisa.it/
- Management Director of ICGDR (International Consortium on Geo-disaster Reduction), https://icgdr.com/
- Associate Editor of Geoenvironmental Disaster Journal, Springer Nature, https://geoenvironmental-disasters.springeropen.com/
- Board Officer for the Italian Chapter of IGS (International Geosynthetics Society), https://www.geosyntheticssociety.org/
- Research topics including: landslide mechanisms, solid–fluid transition, landslide dynamics, regional slope stability, slope erosion, geosynthetics reinforcement, laboratory testing of unsaturated soils, constitutive modeling
- More than 100 international publications (most in peer-reviewed international ISI Journals).

- International Cooperation: UPM Madrid, Université Grenoble, NTNU, Trondheim, Universitè Geneve, Northwestern University, Chicago, USA, University Rijeka Croaztia, Beijing University, China
- Editorial Board Member of International Journals: Canadian Geotechnical Journal, Soils and Foundations, ICE Journal of Geotechnical Engineering, Bulletin of Engineering Geology and the Environment, Geoenvironmental Disasters, International Journal of Geosynthetics and Ground Engineering, Journal of Mountain Science

Prof. Junsheng Yang is the professor of School of Civil Engineering at the Central South University of China. He obtained his PhD from Central South University and completed both his undergraduate and his graduate studies at the Hunan University of Science and Technology, Anhui University of Science and Technology, respectively. He had worked as the visiting scholar in the Pennsylvania State University, USA, and TU Bergakademie Freiberg, Germany. His research areas/interests included tunnel and underground works, rock mechanics and engineering. He has authored and co-authored over two hundred articles in peer-reviewed journals, chapters, and conference proceedings.

Unsupervised Learning of Pavement Distresses from Surface Images

Ahmad Abdelmawla, J. James Yang, and S. Sonny Kim(✉)

School of Environmental, Civil, Agricultural, and Mechanical Engineering, College of Engineering, The University of Georgia, Athens, GA 30602, USA
kims@uga.edu

Abstract. The most common asphalt pavement surface distress is cracking, manifested in various forms, such as transverse, longitudinal, and reflective cracks, governed by different initiation and propagation mechanisms. Being able to timely detect and classify different types of cracks provides critical information for properly maintaining and managing our invaluable road assets. In this paper, a series of classic image processing techniques were applied to pavement surface images to delineate crack patterns. The processed images were projected to a low-dimensional feature space through principal component analysis (PCA). A K-means algorithm is then applied to cluster images in the low-dimensional feature space. The results revealed a meaningful correlation between the crack patterns and the clusters derived.

Keywords: Cluster analysis · Principal component analysis · Image processing · Pavement surface cracking

1 Introduction

Accurate pavement condition assessment of a road network is vital for managing maintenance and rehabilitation (M&R) efforts to ensure its continued functionality. Early detection and repair of pavement defects can prevent further degradation and reduce the overall road M&R expenditure. As such, efficient and timely road inspection is one of the key elements of a successful pavement management system. Nevertheless, periodic road condition surveys tend to be costly and time-consuming if being carried out in the conventional fashion, i.e., by human visual inspection.

Good road conditions are important in supporting the proper functions of a country's economy, and many leading civil engineering organizations across the world, such as the American Society of Civil Engineers (ASCE), continue to emphasize the importance of maintaining the good conditions of the roads and other civil infrastructures. However, the existing situation is not optimistic. According to [1], approximately half of the major roads in the U.S. are currently considered as in poor conditions.

Pavement surface condition assessment has historically been conducted manually by state and/or local transportation agencies, which is still the predominant practice in most developing countries. This resource-demanding and time-consuming process

Y. Liu et al. (Eds.): GeoChina 2021, SUCI, pp. 1–8, 2021.
https://doi.org/10.1007/978-3-030-80316-2_1

has motivated the development of various automatic road condition assessment te niques and systems. With the growing interest in automating the pavement conditi survey and rapid advancement in computer vision and deep learning, the past deca has witnessed an emerging popularity of automatic image-based road surface distre assessment systems. However, many existing automatic image-based systems still su fer from variation in environmental conditions, such as lighting condition, shadow ca by nearby objects, rainy weather, etc., and reliability issues associated with underlyi image-based cracks detection algorithms, ranging from simple pixel level filtering complex deep convolutional neural networks [2].

Automating the detection and classification of pavement surface cracks has been challenging task due to multifaceted crack patterns arising from a multitude of facto underlying different crack initiation and propagation mechanisms, such as traffic load ing spectrums, weather-induced material properties, pavement design, etc. Many loca agencies still rely on the traditional visual inspection method for assessment of thei local road networks.

In this study, we aim to leverage the availability of pavement surface images collectec as part of the pavement survey process and apply unsupervised learning techniques (e.g., PCA and cluster analysis) to cluster similar crack types/patterns. Being able to identify different classes of cracks in an unsupervised fashion helps to understand the local dominant crack types and potential contributing factors, and to infer proper treatment strategies. For conceptual proof of this idea, the surface images collected by a survey vehicle from three local roads in Athens, Georgia were utilized.

2 Literature Review

Automated detection of distresses from pavement images is a challenging problem that has been studied by the computer vision research community for more than three decades. The challenges associated with 2D pavement image-based method include variations in image sources (digital cameras, smartphones, unmanned aerial vehicles, etc.), non-uniformity of cracks, surface texture, lack of sufficient background illumination, and the presence of other features such as joints, among others.

Traditional image analysis approaches rely on hand-crafted features. For example, a novel local binary pattern (LBP) based operator for pavement crack detection was introduced in [3] and a crack detection method using Gabor filter was presented in [4]. An automatic crack detection method referred to as CrackTree is developed in [5]. An integrated model for crack detection and characterization is presented in [6] and a variety of image processing algorithms aimed at detecting and characterizing road surface crack distresses are described in [7].

The recent advancement in machine learning (ML), especially in deep learning (DL) with vision tasks (e.g., object detection and classification), has reignited the research in image-based pavement distress detection and classification. In a recent study, Hoang et al. [8] developed a multiclass support vector machine model optimized by artificial bee colony algorithm to achieve an over 96% overall accuracy in classifying longitudinal crack, transverse crack, alligator crack, and no crack. A review on deep learning applications in pavement image analysis and distress detection was conducted by [9].

Specifically, Zhang et al. [10] developed a crack detection method where the discriminative features are learned from raw image patches using ConvNets. Likewise, [11] proposed a deep architecture of CNNs for detecting concrete cracks. More recently, Maeda et al. [12] evaluated Single Shot MultiBox Detector (SSD) with Inception V2 and MobileNet architectures, for detection eight different road damage types.

However, most previous studies on crack detection and classification relied on handcrafted features or were conducted in supervised learning settings. In contrast, our study focuses on clustering pavement surface images that are congruous with the underlying pavement distress classes in an unsupervised fashion through cluster analysis in a low-dimensional feature space of processed images.

3 Data Set

Pavement surface images (800-by-600) were acquired through a dash camera mounted on a survey vehicle. The dataset used in our experiment consists of 1,125 pavement surface images collected from three local roads in Athens, GA, USA.

4 Analysis

Given the complexity of asphalt pavement images, which are typically endowed with various patterns of micro-texture, nonuniform color arising from the mixture of pavement materials, pavement markings, and shadows casted by nearby utility poles, vegetation, etc., preprocessing of the images is usually required to remove noises and highlight cracks. In this study, a series of image processing techniques, including logarithmic transformation, bilateral filters, Canny edge detector [13], and morphological operations, were applied, resulting in "cleaner" images with more distinct crack patterns. The resultant images were further processed by Principal Component Analysis (PCA) to reduce dimensionality of the feature space, followed by a K mean algorithm [14, 15] to cluster the images in the low-dimensional feature space.

4.1 Image Processing

The interference of the image taking with the dynamic environment often induces unwanted noises. Therefore, it is desirable to apply image preprocessing operations first to remove environmental interference as much as possible. In this study, the open-source computer vision library, OpenCV [16] was utilized, which implemented many popular and commonly used image processing and computer vision algorithms. In our experiment, the images were processed in several steps as described below.

First, the originally captured RGB images were resized from 800-by-600 to 100-by-100 to reduce unnecessary resolution as well as computational footprint and converted to grayscale images since colors do not contribute much to crack features (Fig. 1a). Then, the logarithmic transformation was applied, resulting in the range of darker pixels being expanded while brighter pixels being compressed (Fig. 1b).

Random noise is one of the main problems which may affect the results. Due to pavement physical characteristics, the naturally captured images are more or less affected by random noises (e.g., induced by uneven micro-texture) and/or uniform noise (e.g., a result of shadow), which need to be removed. A bilateral filter was applied to smooth images without losing edge precision. Figure 1c shows the result of the same image after applying the bilateral filtering. Finally, a canny edge detector was applied, followed by morphological operations (erosion and dilation) to highlight crack tracks (Fig. 1d).

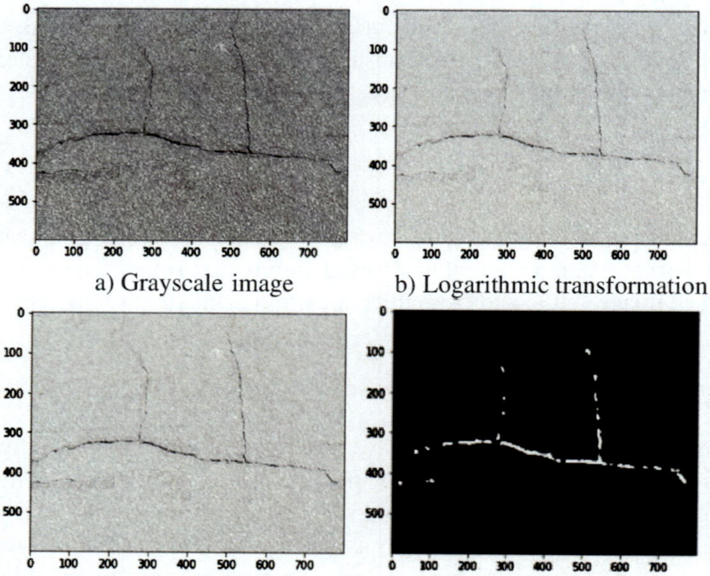

a) Grayscale image b) Logarithmic transformation

c) Bilateral filter d) Canny edge detector & morphological operations

Fig. 1. Image processing.

4.2 Dimension Reduction

With the previously processed images, principal component analysis (PCA) was applied for dimension reduction and feature extraction. In PCA, the high-dimensional features are linearly projected to a low-dimensional space without losing much information. In other words, PCA aims to find a linear subspace of the original feature space that captures the most data variance. In PCA, the principal components (PCs) are created in a descending order of the amount of data variation and all PCs are orthogonal to one another. The eigenvectors of covariance matrix reveal the PC directions while the eigenvalues provide the relative magnitudes of variance explained by corresponding PCs. A limitation of PCA lies in its linearity, which can be addressed by extending the linear kernel to nonlinear kernels through the kernel trick, referred to as kernel PCA [17]. After applying PCA to our dataset, the cumulative variance explained was plotted against the number of components, as shown in Fig. 2. The top 700 PCs, which explained over 80% of the total variance, were retained for cluster analysis.

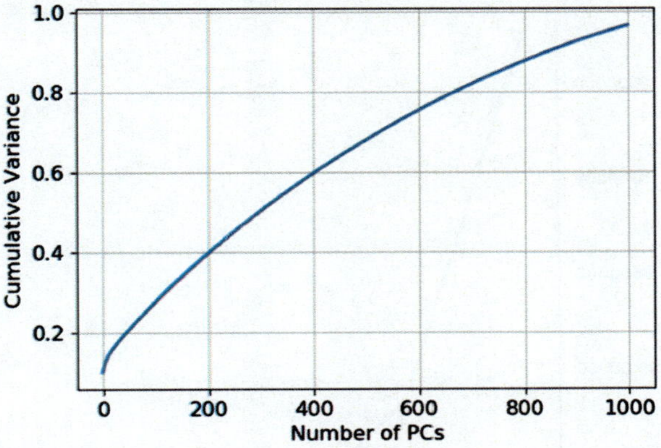

Fig. 2. Cumulative variance explained versus the number of principal components.

4.3 Cluster Analysis

A variety of cluster analysis methods have been developed with K-means (Lloyd 1957; MacQueen 1967) being the most popular one that is flexible and works well in many settings. In K-means, the number of clusters, K, needs to be prespecified. The elbow method has been typically used to determine the K value that corresponds to the point of the maximum curvature in the inertia plot. Given a dataset, the goal of K-means is to find the K clusters with the minimal within-cluster variation. Using the Euclidean distance as the proximity measure, K-means algorithm can be expressed as an optimization problem in Eq. 1.

$$\underset{C_1,\ldots,C_k}{\text{minimize}}\left\{\sum\nolimits_{k=1}^{K}\frac{1}{|C_k|}\sum\nolimits_{i,i'\in C_k}\sum\nolimits_{j=1}^{p}\left(x_{ij}-x_{i'j}\right)^2\right\} \tag{1}$$

where, x_{ij} = the jth feature of observation i; p = the number of features; $|C_k|$ = the number of observations in the kth cluster.

The within-cluster sum-of-squares in Eq. 1 is often referred to as inertia. Applying K-means to our dataset in the reduced feature space, the resulting inertia is plotted against the number of clusters (K) in Fig. 3. Per the elbow method, three clusters (K = 3) were chosen.

For visualization purposes, the cluster results are plotted in the 2D plane formed by the first two principal components (see Fig. 4).

The clusters are indicated by colors in Fig. 4. Cluster 0 (blue) represents multi-directional cracks, Cluster 1 (orange) coincides with longitudinal crack, Cluster 2 (green) contains mainly images without cracks. Exemplar images from each cluster are shown in Fig. 5.

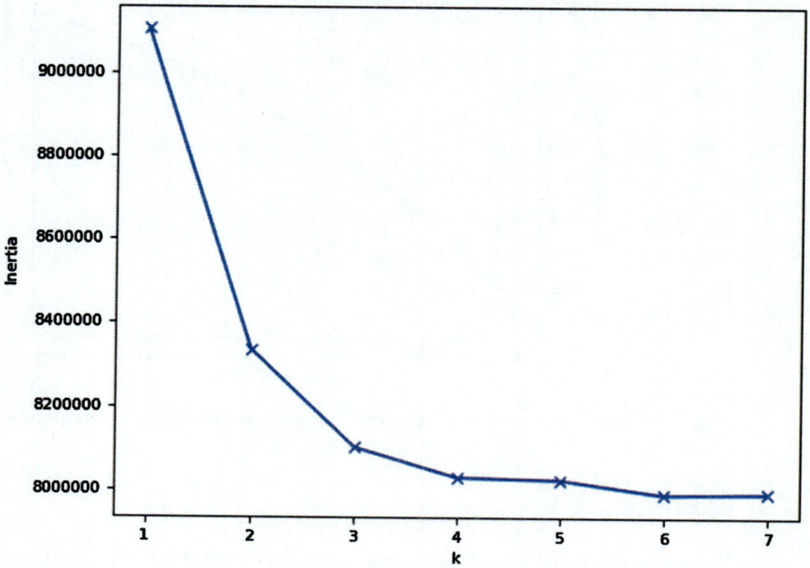

Fig. 3. Inertia versus number of clusters (K).

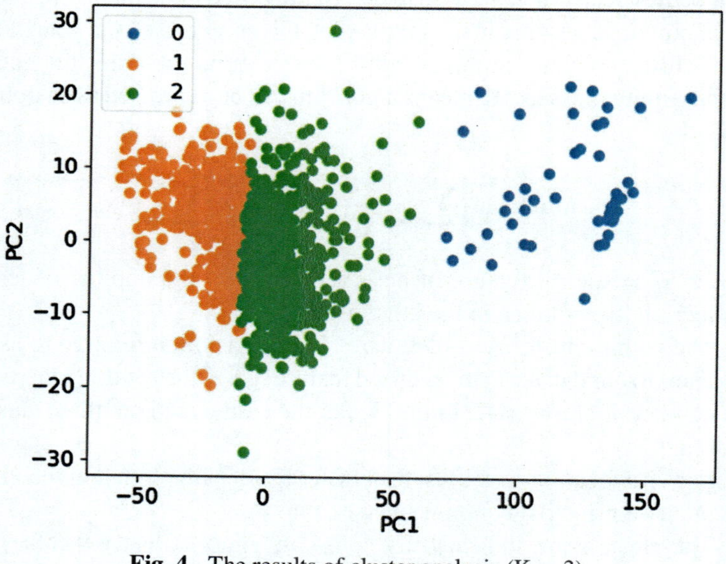

Fig. 4. The results of cluster analysis (K = 3).

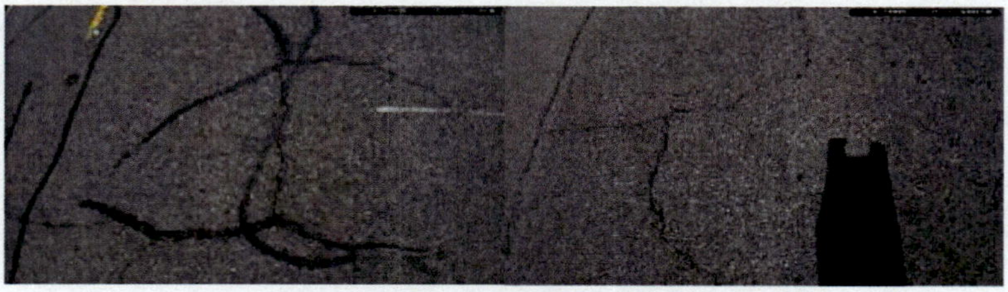

a) Cluster 0 (multi-directional cracks)

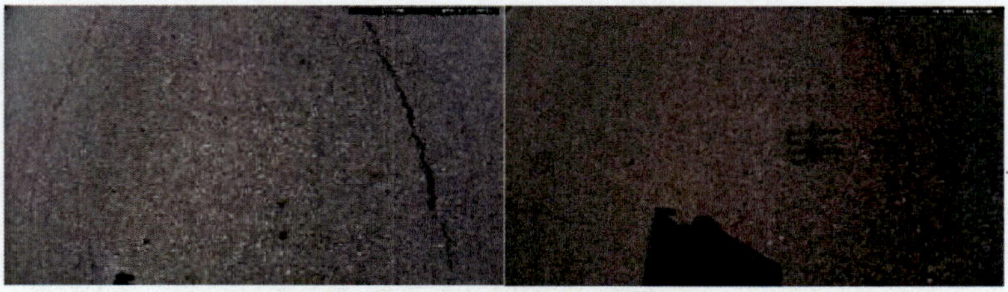

b) Cluster 1 (longitudinal cracks)

c) Cluster 2 (no apparent cracks)

Fig. 5. Exemplar images in each cluster.

5 Conclusions

As a popular nonintrusive technology, research in image-based pavement condition assessment has recently regained momentum by leveraging modern machine learning methods and techniques. This paper presents preliminary results from an ongoing study, demonstrating the use of simple unsupervised machine-learning methods (i.e., PCA for feature extraction and K-means for cluster analysis) for clustering pavement surface distresses. The results revealed a meaningful correlation between derived clusters and crack types. The simple approach could be used to help local jurisdictions understand their predominant pavement distress types and plan for M&R activities. The cluster outcome can also be used as a pre-step to assist with image annotation for developing supervised classification models. Given the limitation of PCA in its linear projection, further studies should consider nonlinearity in feature extraction, such as through kernel PCA, variational autoencoders, etc.

Besides the pavement surface images, modern pavement survey practices also tend to capture subsurface profiles using the ground penetrating radar (GPR) technology. Surface images emphasize functional failure while subsurface GPR scan images can reveal structural failure. For effective diagnosis of pavement structure, surface images and GPR scan images can be jointly used to enrich the feature space through feature engineering techniques to properly capture the complementary aspects of pavement structure.

References

1. AASHTO: Rough Roads Ahead: Fix Them Now or Pay for it Later. Joint Publication of the American Association of State Highway and Transportation Officials and National Transportation Research Group, Washington, DC (2009)
2. Riid, A.; Lõuk, R.; Pihlak, R.; Tepljakov, A.; Vassiljeva, K.: Pavement distress detection with deep learning using the orthoframes acquired by a mobile mapping system. Appl. Sci. **9**, 4829 (2019)
3. Hu, Y., Zhao, C.X.: A novel LBP based methods for pavement crack detection. J. Pattern Recogn. Res. **5**(1), 140–147 (2010)
4. Salman, M., Mathavan, S., Kamal, K., Rahman, M.: Pavement crack detection using the gabor filter. In: Proceedings of IEEE International Conference on Intelligent Transportation Systems, pp. 2039–2044 (2013)
5. Zou, Q., Cao, Y., Li, Q., Mao, Q., Wang, S.: CrackTree: automatic crack detection from pavement images. Pattern Recogn. Lett. **33**(3), 227–238 (2012)
6. Oliveira, H., Correia, P.L.: Automatic road crack detection and characterization. IEEE Trans. Intell. Transp. Syst. **14**(1), 155–168 (2013)
7. Oliveira, H., Correia, P.L.: Crackit-an image processing toolbox for crack detection and characterization. In: Proceedings of IEEE International Conference on Image Processing, pp. 798–802 (2014)
8. Hoang, N.-D., Nguyen, Q.-L., Bui, D.T.: Image processing-based classification of asphalt pavement cracks using support vector machine optimized by artificial bee colony. J. Comput. Civ. Eng. **32**, 04018037 (2018)
9. Gopalakrishnan, K.: Deep Learning in data-driven pavement image analysis and automated distress detection: a review. MDPI J. Data **3**, 28 (2018)
10. Zhang, L., Yang, F., Zhang, D., Zhu, Y.J.: Road crack detection using deep convolutional neural network. In: Image Processing (ICIP), IEEE International Conference, pp. 3708–3712 (2016)
11. Cha, Y.J., Choi, W., Büyüköztürk, O.: Deep learning-based crack damage detection using convolutional neural networks. Comput. Aided Civ. Infrastructure Eng. **32**(5), 361–378 (2017)
12. Maeda, H., Sekimoto, Y., Seto, T., Kashiyama, T., Omata, H.: Road damage detection and classification using deep neural networks with smartphone images. Comput.-Aided Civ. Infrastructure Eng. **33**, 1127 (2018)
13. Canny, J.: A computational approach to edge detection. IEEE Trans. Pattern Anal. Mach. Intell. **8**(6), 679–698 (1986)
14. Lloyd, S.P.: Least squares quantization in PCM. Technical Report RR-5497, Bell Lab, September 1957. Google Scholar
15. MacQueen, J.B.: Some methods for classification and analysis of multivariate observations. In: Le Cam, L.M., Neyman, J. (eds.) Proceedings of the Fifth Berkeley Symposium on Mathematical Statistics and Probability, vol. 1, pp. 281–297. University of California Press, California
16. OpenCV: Open Source Computer Vision Library (2021)
17. Schölkopf, B., Smola, A., Müller, K.-R.: Kernel principal component analysis. In: Gerstner, W., Germond, A., Hasler, M., Nicoud, J.-D. (eds.) ICANN 1997. LNCS, vol. 1327, pp. 583–588. Springer, Heidelberg (1997). https://doi.org/10.1007/BFb0020217

Physical-Mechanical Properties of Sandstones After Exposure to High Temperature

Gang Wu[1], Hong Sun[1(✉)], Gusheng Tong[2], and Wang Yu[3]

[1] School of Naval Architecture, Ocean and Civil Engineering, Shanghai Jiao Tong University, Shanghai 200240, China
sunhong@sjtu.edu.cn

[2] School of Civil and Architectural Engineering, East China Jiaotong University, Nanchang 330013, Jiangxi, China

[3] Institute for Materials Research, School of Computing, Science and Engineering, University of Salford, Salford M5 4WT, UK

Abstract. Studying the problems of rock engineering under high temperatures has become a new research direction for rock mechanics. The physical-mechanical properties of rocks under high temperatures circumstances have a great interest in many of engineering applications, such as deep geological repositories for heat-generating radioactive wastes, oil/gas recovery enhancement, underground gasification of coals, exploitation of geothermal resources and the protection of buildings against fire or restoration after exposure to fires. In this paper, the physical and mechanical properties of Jiaozuo sandstone after experiencing heat to high temperatures are studied. The temperature varies in a range from 20 °C to 1200 °C. The studied physical properties include the changes of samples' apparent shape, volume, mass, density, and velocity of longitudinal and shear waves through these samples. The studied mechanical properties include peak stress, peak strain, and Young's modulus under uniaxial compression state. The deterioration mechanism of sandstones exposed to high temperatures is briefly discussed by detecting mineral composition. The test results show that high temperature leads to the changes of apparent shape for sandstone, the temperature does not obviously affect the mechanical properties, and heating reinforces some mechanical properties of sandstone in the temperature ranging from room temperature to 400 °C. When the temperature is above 400 °C, the mechanical properties of sandstone present deterioration with the increase of temperature, the peak stress and Young's modulus of sandstone decrease in different extents. The peak strain of sandstone increases by a big margin before 800 °C. The deformation of sandstone generally increases with the rising of temperature. From 600 °C to 1200 °C, clear brittle-plastic transition appears in sandstone. Sandstone presents plastic failure characteristics after 1200 °C. Thermal stress, changes in mineral formation, and microstructure due to temperature result in changes in mechanical properties and degradation of sandstone.

Keywords: Rock mechanics · High temperature · Sandstone · Physical and mechanical characteristics

Y. Liu et al. (Eds.): GeoChina 2021, SUCI, pp. 9–18, 2021.
https://doi.org/10.1007/978-3-030-80316-2_2

1 Introduction

The thermo-mechanical properties of rocks are one of the important problems in geotechnical engineering, such as exploitation of geothermal resources, underground gasification of coals, deep geological disposal for high radioactive wastes (Laloui et al. 2008), oil/gas recovery enhancement, and the protection or restoration of buildings against fire after exposure to fires.

The physical-mechanical behaviours of rocks changed subject to thermal loads. David et al. (1999) presented the cracks of granites developed completely differently exposed to high temperatures. The cracks due to mechanical loading were manifested by anisotropy, while by isotropy due to thermal loading. Hajpal (2002) studied the effects of high temperatures on sandstones in Germany and Hungary and reported that ingredients, porosities, fractures, and wave transportation varied from 150 °C to 900 °C. Chen et al. (2017) presented that the uniaxial compressive strength (UCS) and Young's modulus of the granite decreased with an increase in temperature up to 800 °C. Zhang et al. (2019) presented that the sandstone had a threshold temperature near 400 °C. New cracks developed gradually, while the original cracks expanded rapidly above the threshold temperature. Liu and Xu (2015) studied granite and sandstone and reported that the brittle-ductile transition critical temperatures were observed in their stress–strain curves. Zhang et al. (2020) reported the brittle-ductile transition of the mudstone occurred at about 800 °C. The high temperature has a strong influence on the physical-mechanical behaviours of rocks. Therefore, it is necessary to study the variation and deterioration mechanism of rocks after exposure to high temperature.

However, the temperature measured in previous studies were all below 900 °C. The physical and mechanical properties of stones can not be investigated in a wider temperature range. So the transformation of mineral crystal form and the appearance of macro cracks were not observed due to higher temperature.

The work aims to study the physical-mechanical properties of sandstones after exposure to a high temperature by physical measurements and unconfined compression tests, to analyse the deterioration mechanism of mechanical parameters exposed to high temperature. The results provide a scientific basis for predicting and evaluating the stability and safety of rock mass engineering after high temperatures and for the repair and reinforcement of rock mass engineering after fires.

2 Samples and Test Procedure

2.1 Test Samples

Sandstone samples were taken from a coal mining site in Jiaozuo city, Henan province, China. Sandstone was composed of quartz, feldspar, calcite siderite, mica, and charcoal. Its average density at room temperature is 2.65 g/cm^3. The specimen's cylinder was made with 50 mm in diameter and 100 mm in height without micro-cracks and macro-pores. Ultrasonic longitudinal and transverse wave velocities of specimens were obtained by an NM-4B ultrasonic test device. The specimens were selected with similar ultrasonic wave velocities.

2.2 Test Procedure

Based on the existing literatures and the comprehensive procedure influenced by all temperatures, the experiments were carried out at different temperatures, which were 20 °C, 100 °C, 150 °C, 200 °C, 400 °C, 600 °C, 800 °C, 1000 °C, and 1200 °C. The quantities of specimens at different temperatures are shown in Table 1. Each specimen was placed in an SX2-8-16 electronically controlled heating oven manufactured by Shanghai Zufa Industrial Ltd. The temperature was increased at a rate of 5 °C per min until the target temperature was reached. It was maintained for a minimum of 2 h, and then the specimens were cooled to the room temperature. The specimens were taken out of the oven for tests. The physical parameters and the ultrasonic wave velocities were measured. The uniaxial compression experiments were carried out by an RMT-150B rock mechanical testing system from the Institute of Rock and Soil Mechanics, Chinese Academy of Sciences. All tests were strain-controlled at 0.0005 mm/s for 20 °C–400 °C, 0.001 mm/s for 600 °C–1000 °C, and 0.005 mm/s for 1200 °C. All tests were performed according to the Chinese standard (ECPRC 2013).

Table 1. Quantities of sandstone specimens at different temperatures

Temperature °C	Number	Quantity	Temperature °C	Number	Quantity
20	A-1–A-5	5	600	F-1–F-11	11
100	B-1–B-5	5	800	G-1–G-15	15
150	C-1–C-5	5	1000	H-1–H-11	11
200	D-1–D-5	5	1200	I-1–I-7	7
400	E-1–E-11	11			

3 Test Results

3.1 Physical Properties

The temperature can change the physical properties of sandstones, especially the appearance and colors of specimens significantly changed above 400 °C, shown as Fig. 1. Below 400 °C, their colors were dark grey. They became grey, and stripes appeared at 600 °C, pale grey and reddish-brown stripes appeared at 800°C, taupe and there were apparent cracks at 1000 °C, large cracks occurred along the stripes and precipitates in the cracks appeared at 1200 °C.

The masses and densities of the specimens decreased with the increasing temperature, shown in Figs. 2 and 3 in which m is the mass, ρ is the density and Δ is a symbol of variation. The rates of mass loss and density were different at different temperatures. The mass and density losses were slight below 200 °C, significantly above 400 °C. However, the mass losses ratio almost stopped above 1000 °C, and the density reduced abruptly at 1200 °C. The mass losses were subject to evaporation of pore water, loss of structural water, and changes of some minerals due to heating; simultaneously, the

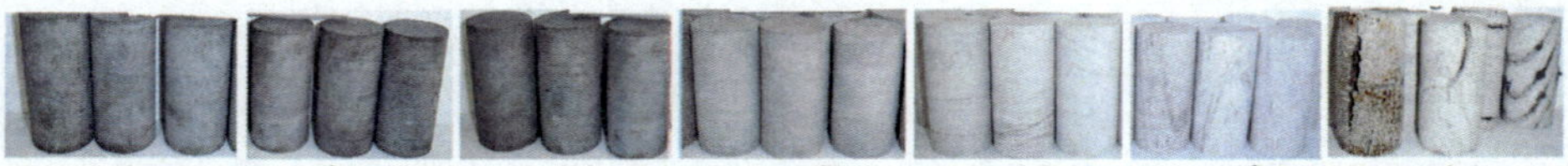

Fig. 1. The photos of specimens at different temperatures: a 20 °C; b 200 °C; c 400 °C; d 600 °C; e 800 °C; f 1000 °C; g 1200 °C

volume of specimen increased, resulting in the decreased density, especially the micro-cracks appeared at 1200 °C.

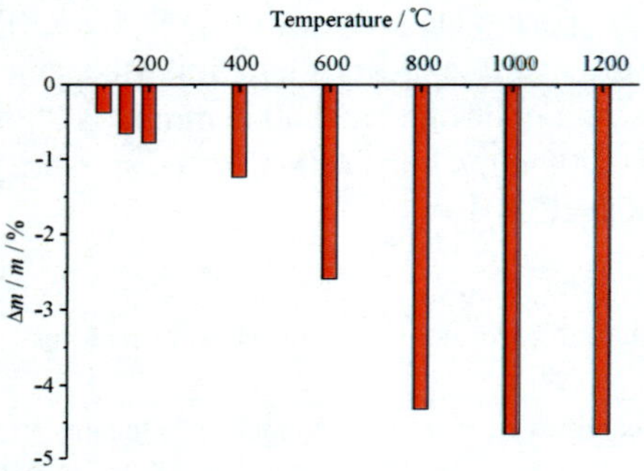

Fig. 2. The change ratio of average mass for sandstone before and after heating

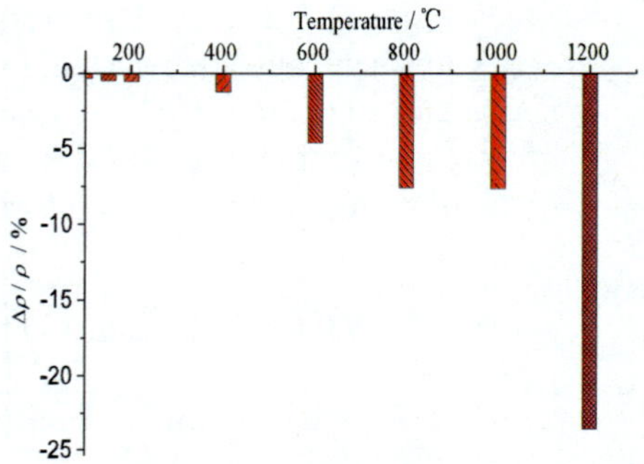

Fig. 3. The change ratio of average density for sandstone before and after heating.

3.2 Longitudinal and Transverse Wave Velocities

Figure 4 presents the longitudinal and transverse wave velocities of sandstone specimens at different temperatures in which V_p is the velocity of the longitudinal wave, V_s is the velocity of the transverse wave, Δ is a symbol of variation. The transverse wave velocities were larger than longitudinal ones. They had a similar variation with the increasing temperature. Below 200 °C, slight changes of both of them were observed. However, from 400 °C to 800 °C, they significantly decreased. For example, longitudinal and transverse wave velocities respectively reduced by 38.86%, 35.29% at 800 °C. From 1000 °C to 1200 °C, they decreased with a small degree.

The velocity ratio of longitudinal to transverse wave is an important parameter that indicates the relationship between compression and shearing loading. It is relative to the elastic parameters of materials (Chen and Huang 2001). It decreased with the increasing temperature below 800 °C, but it increased from 800 °C to 1200 °C, as shown in Fig. 5. It indicated that temperature changed their response to longitudinal and transverse waves.

3.3 Dynamic Young's Modulus

Based on the elastic mechanical theory, the dynamic elastic modulus can be expressed using the velocities of the longitudinal and transverse waves as follows (Jaeger et al. 2007):

$$E_d = \rho V_s^2(3V_p^2 - 4V_s^2)/(V_p^2 - V_s^2) \tag{1}$$

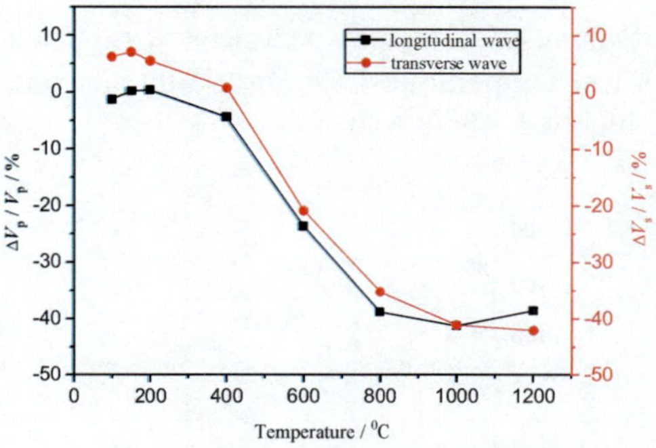

Fig. 4. The changes of wave velocities for sandstone at different temperatures

where E_d is the dynamic Young's modulus (GPa), ρ is the density (kg/m^3), V_p is the velocity of the longitudinal wave (m/s), and V_s is the velocity of the transverse wave (m/s).

Figure 6 shows the Young's moduli and dynamic Young's moduli before and after heating at different temperatures. The Young's moduli, which were the slopes of the approximate straight line of the stress-stain curve by uniaxial compression tests, were lower than dynamic ones, which was in accordance with Wang (1997). After heating, there were many differences at different temperatures. The Young's moduli and dynamic ones after heating significantly decreased in the range from 400 °C to 800 °C. The average dynamic Young's modulus reached 36 GPa at 400 °C, 12 GPa at 800 °C. When the temperature was above 1000 °C, dynamic Young's modulus after heating changed slightly. It reached approximately 11 GPa, while Young's modulus was 2 GPa at 1200 °C.

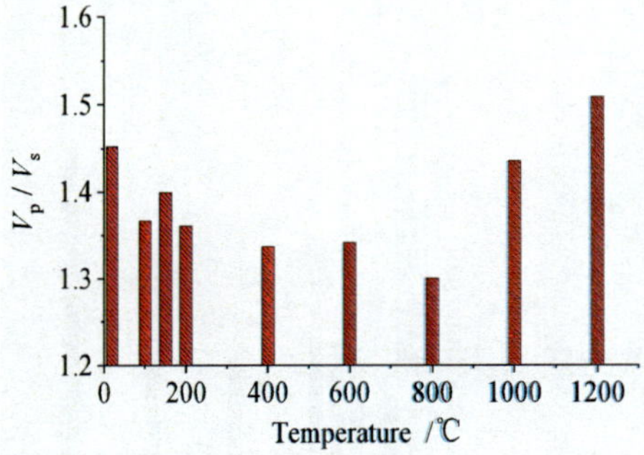

Fig. 5. The average ratio of wave velocity for sandstone at different temperatures

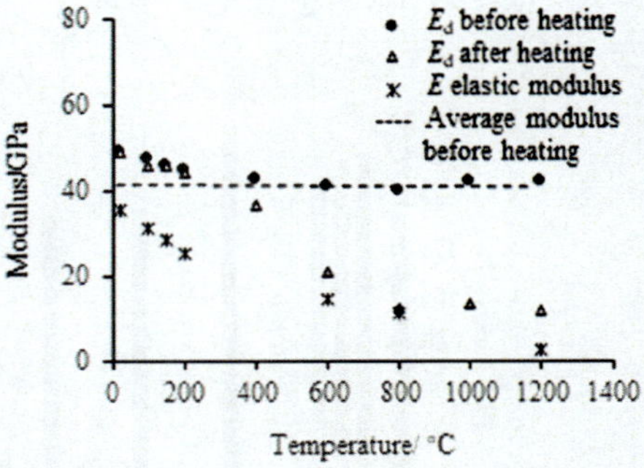

Fig. 6. The Young's moduli and dynamic Young's moduli before and after heating

3.4 Mechanical Properties by Unconfined Compression Test

The typical stress-strain curves after heating at different temperatures were shown in Fig. 7. The stress was approximately linear to strain at initial loading, and the post-peak curves were characterized by a sudden loss in strength. Above 600 °C, the curves were more gentle. It indicated that the specimens at high temperatures were more ductile than those of lower temperatures. The brittle-ductile transition occurred at a threshold temperature, which was about 600 °C.

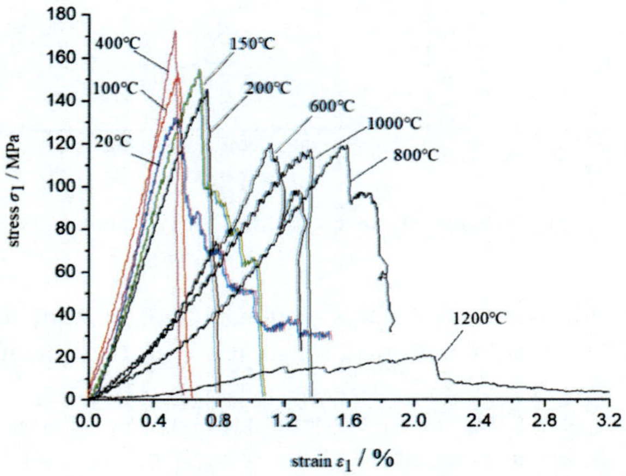

Fig. 7. Typical stress-strain curves of sandstones after exposure to different temperatures

The peak stresses were different at different temperatures, shown in Fig. 8. The highest one was 173 MPa after exposure to 400 °C. Above 600 °C, it decreased with the increasing temperature. Especially the strengths significantly reduced to below 20 MPa at 1200 °C.

The strain at the peak stress changed with the increasing temperature, shown in Fig. 9. The strain was larger than 10.5% above 600 °C. Especially the maximum strain was 14.5% at 800 °C.

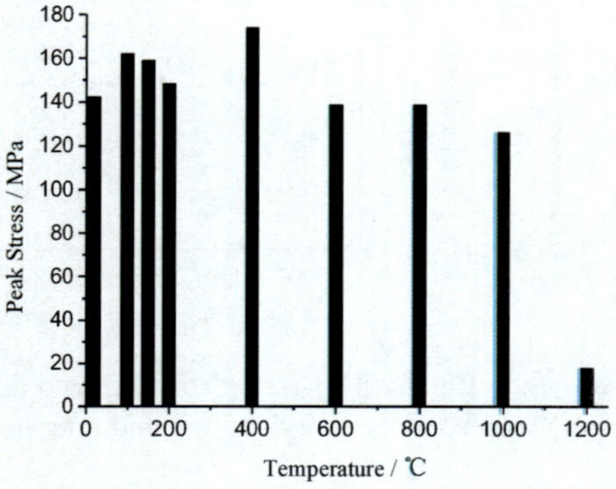

Fig. 8. The peak stress at different temperatures

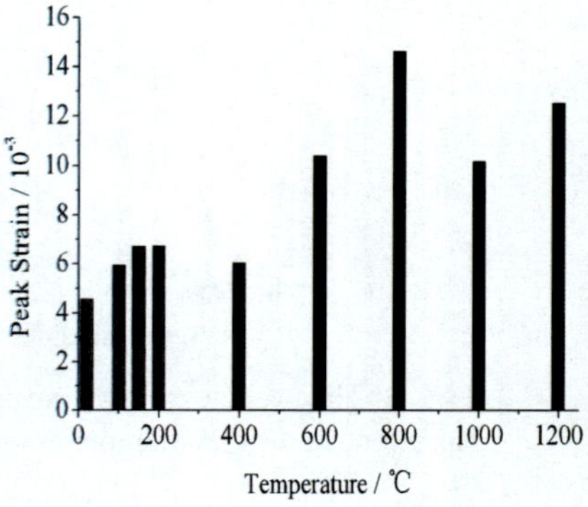

Fig. 9. The peak strain at different temperatures

4 Discussion

Although test results were deviated and dispersed which were due to the inhomogeneity in natural structure, end effect, and specimen preparation, the influence of temperature on the physical-mechanical properties of sandstones could be investigated, and the high-temperature deterioration could be discussed in terms of thermal stress mechanism and material physical properties.

Sandstones were composed of various minerals with different particle sizes, thermal expansion rates, and thermal anisotropy and elastic characteristics. With the increasing temperature, the thermal stress (tensile or compressive) between or within particles appeared due to the incompatible thermal expansion across particle boundaries. With the increase in thermal stress, microfractures inside sandstones occurred, subsequently developed into big macroscopic cracks. The deterioration of the mechanical properties of sandstones was observed at a macro-scale.

4.1 The Decrease of Elastic Wave Velocities at High Temperature

Mineral composition, pores, and microfractures have strong influences on the wave transport velocities through rocks (Luo and Wang 2011; Wu et al. 2012). Under higher temperature, the mineral composition of sandstones changed. Sandstones mainly consisted of quartzes and sericites. Under 600 °C, all of the sandstone specimens contained feldspars, dolomites, calcites, and carbons. Their content ranged from 0.5% to 3%. Above 800 °C, calcites and carbons disappeared. Above 1000 °C, dolomites disappeared. Above 1200 °C, feldspars were not measured, simultaneously, iron was detected (He 2006). The original structure was destroyed by melting and recrystallizing of mineral composition when heated to a high temperature so that the velocities of wave transporting through sandstones decreased. On the other hand, the denser the interaction between particles, the faster the wave transportation. After heated, the microfractures appeared inside the sandstone specimens, the interaction between particles was reduced, so that the transporting resistance of elastic wave increased, and the wave velocity decreased. There would be a decrease in longitudinal and transverse wave velocities with the increasing temperature.

4.2 The Deterioration of Mechanical Parameters

From 20 °C to 100 °C, there were increases in strength and peak strain due to water evaporation inside sandstones. From 100 °C to 200 °C, various minerals adjusted, some microfractures developed to larger fracture, resulting in a decrease of the peak strength but an increase in peak strain. Meanwhile, Young's modulus decreased. Therefore, the mechanical performance of sandstones was deteriorated due to the higher temperature.

From 200 °C to 400 °C, some minerals of sandstones were melted to fill into some fractures, the number of fractures decreased. There was an improvement in mechanical properties. Peak strength, peak strain, and Young's modulus increased. Above 600 °C, mineral compositions and the internal structure of sandstone changed due to the high temperatures. For example, quartzes were transformed from α-state to β-state although changed back to α-state after cooling. The process of phase transformation affected the association of mineral particles. At 600 °C, the elastic parameters were significantly lower than that at 400 °C because the elasticity of sericites changed. Above 800 °C, the mineral composition changes included the loss of structural water, recrystallization and phase change, etc. The deterioration in mechanical properties was observed. Above 1200 °C, visible cracks occurred, and dark brown solid (vitreous) substance appeared in these cracks (Fig. 8), the natural structures of specimens were completely destroyed.

According to the statistical analysis (He 2006; Zhang 2005), below 400 °C, the mass of sandstone gradually decreases with the increase of the temperature, but the change rate of mass was less than 1.50%. The volume decreased slightly, and its change rate was less than 0.50%. From 600 °C to 800 °C, the mass of sandstone decreased obviously with reduction rates of 2.59% and 4.32%. After that, the change of sandstone mass tended to slow down. Above 400 °C, the sandstone volumes increased. The average growth rates of sandstones were respectively 1.14%, 2.41%, and 3.57% at 600 °C, 800 °C, and 1000 °C. The average volume growth rate reached 24.06%. The strength of sandstone with large mass loss and noticeable stripes decreased significantly.

The variation laws of physical and mechanical parameters of sandstones after high temperature are consistent with the meso observation results in reference (Zhang et al. 2005).

In addition, the mechanical properties of rocks after exposure to high temperatures are dependent on temperature path and closely related to temperature change rates, temperature history, and stress environment. Therefore, the deterioration of rock due to high temperature is a thermal-mechanical coupling problem. It is necessary to develop high-temperature rock testing technology and comprehensively use rock mechanics, thermodynamics, geochemistry, and mineralogy theories to fully reveal the high-temperature deterioration mechanism of rocks. It is an important research direction in the future.

5 Conclusions

The physical and mechanical properties of sandstone after exposure to high temperature were investigated. The following conclusions can be drawn:

1. The mechanical properties are damaged by the increasing temperature with different degrees. The damage procedure can be divided into three stages: slight damage stage from 20 °C to 400 °C; damage stage from 400 °C to 800 °C; failure stage above 800 °C.
2. The appearance and density of sandstone significantly changed, subjected to heating above 400 °C. The variations of wave velocities, the Young's modulus, and dynamic modulus on average decreased as temperature increased. The peak stress decreased significantly decreased in the range from 400 °C to 800 °C. At 800 °C, specimens became white grey, stripes appeared, and mass decreased. At 1200 °C, visible cracks appeared. The peak stress drops significantly when heated up to 1200 °C.
3. The variations of wave velocities are different at different temperatures. Below 200 °C, the wave velocities increased with the increasing temperature, but above 400 °C, they decreased as temperature increased. There was a temperature threshold from 600 °C to 800 °C, and the velocities dropped when it was higher than the threshold.

4. The deterioration of sandstone after exposure to high temperature is due to the loss of adsorbed water, the transformation of mineral crystal form, recrystallization, and the escape of structural water.

Acknowledgement. The research described in this paper was financially supported by the Natural Science Foundation of China under Grant No. 41572255 and the Natural Science Foundation of Shanghai Grant No. 15ZR1423300.

References

Chen, Y., Huang, T.F.: Rock Physics. Peking University Press, Beijing (2001). (in Chinese)

Chen, Y.L., Wang, S.R., Ni, J., et al.: An experimental study of the mechanical properties of granite after high temperature exposure based on mineral characteristics. Eng. Geol. **220**, 234–242 (2017)

Electricity Council of the People's Republic of China (ECPRC, 2013): Standard for Test Methods of Engineering Rock Mass (GB/T 50266-2013). China Planning Press, Beijing, China (2013). (in Chinese)

David, C., Menéndez, B., Darot, M.: Influence of stress-induced and thermal cracking on physical properties and microstructure of La Peyratte granite. Int. J. Rock Mech. Min. Sci. **36**(4), 433–448 (1999)

Hajpál, M.: Changes in sandstone of historical monuments exposed to fir e or high temperature. Fire Technol. **38**(4), 373–382 (2002)

He, G.L.: Research on the elastoplastic disturbed state concept model of rock and high temperature, freeze-thaw tests of rock. Master. D. thesis, Shanghai Jiaotong University, Shanghai (2006). (in Chinese)

He, G.L., Wu, G., Huang X.C., et al.: Experimental study on ultrasonic properties of sandstone before and after high temperature. Rock Soil Mech. **28**(4), 779–784 (2007). (in Chinese)

Jaeger, J.C., Cook, N.G.W.: Zimmerman, R.W.: Fundamentals of Rock Mechanics, 4th edn. Chapman & Hall, London (2007)

Luo, J.A., Wang, L.G.: High-temperature mechanical properties of mudstone in the process of underground coal gasification. Rock Mech. Rock Eng. **44**, 749–754 (2011)

Liu, S., Xu, J.Y.: An experimental study on the physico-mechanical properties of two post-high-temperature rocks. Eng. Geol. **185**, 63–70 (2015)

Wang, R.J.: The Acoustic Classification of Rocks and Study of Dynamic-Elastic Mechanical Parameters of Rocks. Geology Press, Beijing (1997). (in Chinese)

Wu, G., Wang, D.Y., Zhai S.T., et al.: Test research on mechanical properties of marble under high temperature. Chin. J. Rock Mech. Eng. **31**(6), 1237–1244 (2012). (in Chinese)

Zhang, L.Y., Mao, X.B., Lu, A.H.: Experimental study on the mechanical properties of rocks at high temperature. Sci. China Ser. E Technol. Sci. **52**(3), 641–646 (2009)

Laloui, L., Francois, B., Nuth, M., et al.: A thermo-hydro-mechanical stress-strain framework for modelling the performance of clay barriers in deep geological repositories for radioactive waste. In: Toll, D., et al. (eds.) Unsaturated Soils: Advances in Geo-Engineering, pp. 63–80. Taylor & Francis Group, London (2008)

Zhang, L.Y., Mao, X.B., Li, M., et al.: Brittle-ductile transition of mudstone in coal measure rock strata under high temperature. Int. J. Geomech. **20**(1), 04019149 (2020)

Zhang, L.: Experimental research and disturbed state concept analysis on physical and mechanical properties of post high-temperature sandstone. Master. D. thesis. Shanghai Jiaotong University, Shanghai (2005). (in Chinese)

Zhang, Y., Zhang, X., Zhao, Y.S.: Process of sandstone thermal cracking. Chin. J. Geophys. **48**(3), 656–659 (2005). (in Chinese)

Zhang, W.Q., Sun Q., Zhu Y.M., Guo, W.: Experimental study on response characteristics of micro–macroscopic performance of red sandstone after high-temperature treatment. J. Therm. Anal. Calorim. **136**(5), 1935–1945 (2019)

Liquefaction Properties of Two Types of Sandy Soil Specimens with Different Compaction Thickness

Tomoko Sasaki[1(✉)], Shima Kawamura[1], and Junichi Koseki[2]

[1] Graduate School of Engineering, Muroran Institute of Technology, 27-1, Mizumoto-chou, Muroran 050-8585, Hokkaido, Japan
sasakit@earth-prime.co.jp

[2] Department of Civil Engineering, The University of Tokyo, 7-3-1 Hongo, Bunkyo-ku, Tokyo 113-8656, Japan

Abstract. It has been pointed out in laboratory element tests that the liquefaction properties of sandy soils are influenced by differences in specimen preparation methods. In practice, the moist-tamping method has been mostly used in laboratory tests because it allows grain size distribution and moisture content to be widely selected. In this study, a series of cyclic undrained triaxial tests was conducted under different conditions of grain size distribution and compaction thickness to reveal the effects of specimen preparation methods on liquefaction properties of two types of sandy soils. The specimens were prepared using the moist-tamping method. The number of compaction layers were either 1, 4, or 10. The results showed that the liquefaction resistance was influenced by compaction thickness, and its tendency changed depending on the grain size distribution and degree of compaction, D_c. In addition, it was found that a difference in the initial moisture content had an influence on deformation properties during cyclic loadings. In particular, the deformation properties of the specimens compacted under lower moisture content was similar to that of loose specimens. Furthermore, in the case of well-graded sand, there was no significant difference in liquefaction resistance between specimens of 10 and 4 layers; however, the liquefaction properties of a 1 layer specimen were significantly different from those of other compaction layers.

1 Introduction

In laboratory tests, the specimen preparation method influences the liquefaction properties of sands (Tatsuoka et al. 1986). Although there are various methods of specimen preparation, the moist-tamping method is especially versatile, as it allows for a wide range of grain size distributions and moisture contents and has high reproducibility.

In general, in specimen preparation by the moist-tamping method, soils are compacted in several batches. However, the specimen becomes non-uniformity during the compaction processes. As a result, the test results might have problems (Sasaki et al. 2020).

Figure 1 shows an example of situation during cyclic loadings for cyclic undrained triaxial test. The specimen in this figure was prepared by the moist-tamping method

Y. Liu et al. (Eds.): GeoChina 2021, SUCI, pp. 19–30, 2021.
https://doi.org/10.1007/978-3-030-80316-2_3

using Toyoura sand, and the number of compaction layers was 4. The specimen in the figure shows the situations around the peak of axial strain in the extension side reached to the initial liquefaction. In the observation of the surface of the specimen, the uneven parts are confirmed for the upper part of the boundary of the layers.

Figure 2 shows variation in density in a specimen of Toyoura sand. The density was measured by the caliper method after cutting the demolded frozen specimen with a high-speed cutter. The broken line in the figure indicates the boundary between the compacted 4 layers of the specimen. The figure also shows that the upper parts are dense and the lower parts are loose, although they are the same thickness of layer. In this way, the moist-tamping method is characterized in that change in density is induced within the same layer in the specimen.

There are many unclear points about the effect of non-uniformity of the specimen on its liquefaction properties when using the moist-tamping method. Therefore, in this study, a series of cyclic undrained triaxial tests was conducted to investigate the effects of the specimen preparation method on the liquefaction properties of two types of sandy soils.

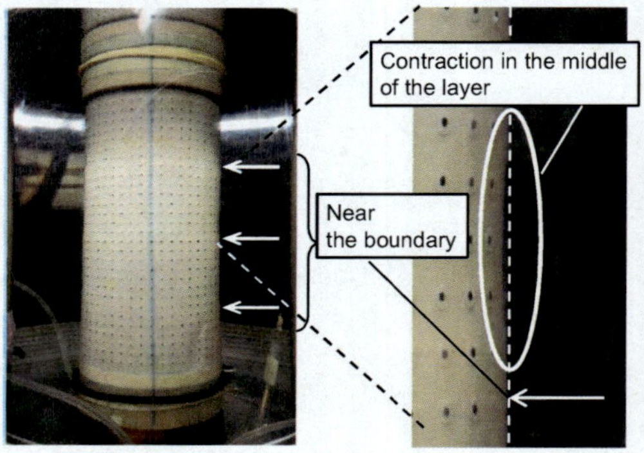

Fig. 1. Specimen during loading (4 layers, CSR = 0.274, double amplitude axial strain is around 5%, near extension peak)

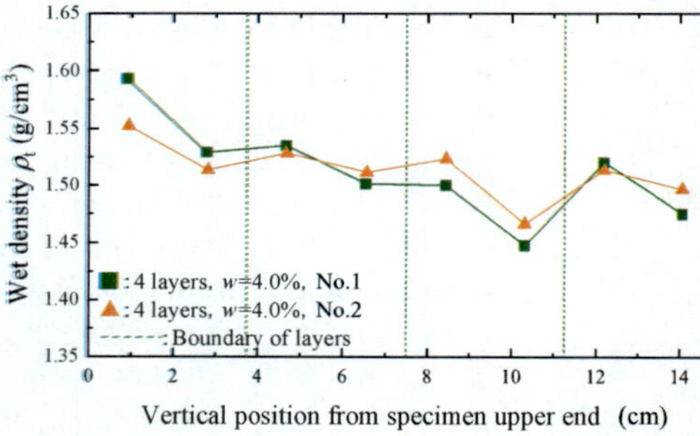

Fig. 2. Local density within the specimen

2 Test Materials

As the test materials, Toyoura sand and well-graded sand were used. The well-graded sand was used for the construction of embankments. Table 1 shows the physical properties of the test materials. Their grain size distributions and compaction curves are shown in Fig. 3 and Fig. 4, respectively. Toyoura sand is a clean one, with very little fines content. The well-graded sand has a wider grain size distribution than that of Toyoura sand.

As shown in Fig. 4, the peak of the compaction curve for Toyoura sand is not clear. This is due to that the effect of compaction is low. On the other hand, the peak of the curve for well-graded sand clearly appears due to that the effect of compaction is high.

The well-graded sand for the cyclic undrained triaxial test was adjusted to the maximum grain size of 4.75 mm in consideration of the specimen diameter of 75 mm.

3 Specimen Preparation and Test Method

The specimens were 75 mm in diameter and 150 mm in height. The specimen preparation method was the moist-tamping method (MT).

Table 1. Index properties of test materials

Parameter	Toyoura sand	well-graded sand	Test method
ρ_s (g/cm^3)	2.651	2.719	JIS A 1202
D_{max} (mm)	0.425	19	JIS A 1204
F_c (%)	0.2	8.9	〃
D_{50} (mm)	0.16	1.3	〃
U_c	1.6	17	〃
e_{max}	0.956	-	JIS A 1224
e_{min}	0.591	-	〃
ρ_{dmax} (g/cm^3)	1.522	1.956	JIS A 1210
w_{opt} (%)	18.2	12.2	〃

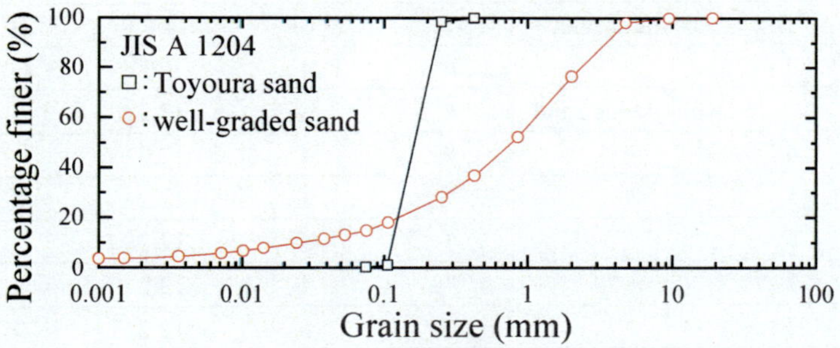

Fig. 3. Grain size distributions of test materials

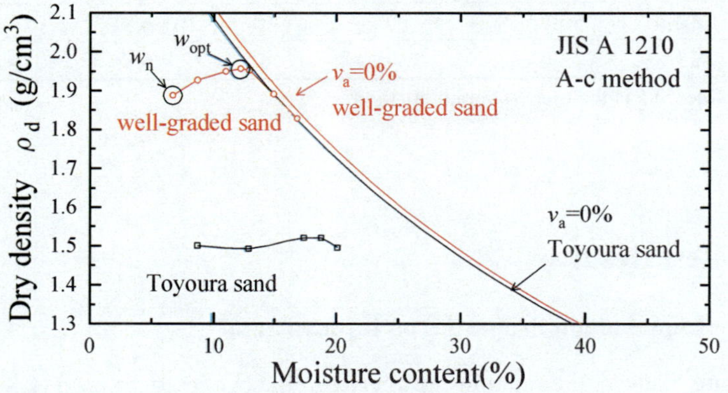

Fig. 4. Compaction curves of test materials

Table 2 shows the condition of test specimens. There were either 1, 4, or 10 compaction layers. The specimens of Toyoura sand were compacted with an initial moisture content of 4.0%. The specimens of well-graded sand were compacted with an initial moisture content of either the optimum content (w_{opt}) of 12.2% or the natural content (w_n) of 6.8%. This is because in practice, the optimum content and the natural content are often used as the initial moisture content.

In the cyclic undrained triaxial test, the specimens were saturated (B-value $\geq$ 0.95) using the double vacuum method. The specimens were then consolidated under an effective confining pressure of 100 kN/m^2.

In cyclic loadings, the deviator stress was kept constant until the double amplitude axial strain (*DA*) reached to 5%. The Toyoura sand specimens were loaded under the strain-controlled conditions with a loading rate of 0.1%/min. The well-graded sand specimens were loaded under stress-controlled conditions with a frequency of 0.01 Hz. The cyclic undrained triaxial tests were conducted under two loading conditions; however, the results did not differ between the two conditions. Other testing procedures were based on the "Method for cyclic undrained triaxial test on soils" (JGS 0541-2009).

Table 2. Conditions of test specimens

No.	Material	Method	Layer	w (%)	D_r (%)	D_c (%)
1	Toyoura sand	MT	10	4.0	60	100
2	〃	〃	4	4.0	60	100
3	well-graded sand	〃	10	12.2	-	95
4	〃	〃	4	12.2	-	95
5	〃	〃	1	12.2	-	95
6	〃	〃	10	12.2	-	90
7	〃	〃	4	12.2	-	90
8	〃	〃	1	12.2	-	90
9	〃	〃	10	12.2	-	85
10	〃	〃	4	12.2	-	85
11	〃	〃	1	12.2	-	85
12	〃	〃	10	6.8	-	90
13	〃	〃	4	6.8	-	90

Note: MT is the moist-tamping method

4 Test Results

4.1 Liquefaction Properties of Toyoura Sand

Figure 5 shows the relationships between the cyclic stress ratio (CSR) and the number of loading cycles (N_c) required until *DA* of 5% for Toyoura sand. It is found from the figure that the liquefaction curve for the specimen with 4 layers is lower than that for the specimen with 10 layers.

Figures 6 shows the data for around 20 loading cycles (see Fig. 5) in terms of the relationship between the residual pore water pressure ratio $\Delta u/\sigma_c'$ and the double amplitude axial strain *DA* normalized by $DA = 5\%$. Also, Fig. 7 shows the relationship between *DA* normalized by $DA = 5\%$ and N_c required until $DA = 5\%$. These are the data for around 20 loading cycles as Fig. 6. From Fig. 6, it is found that the difference in the number of compaction layers has a little effect on the development of residual pore water pressure ratio.

On the other hand, Fig. 7 shows that the axial strain of the specimen with 4 layers develops more gradually than that with 10 layers. A similar behavior to specimen with 10 layers, which axial strain increases sharply during cyclic loadings, is generally confirmed for loose specimen.

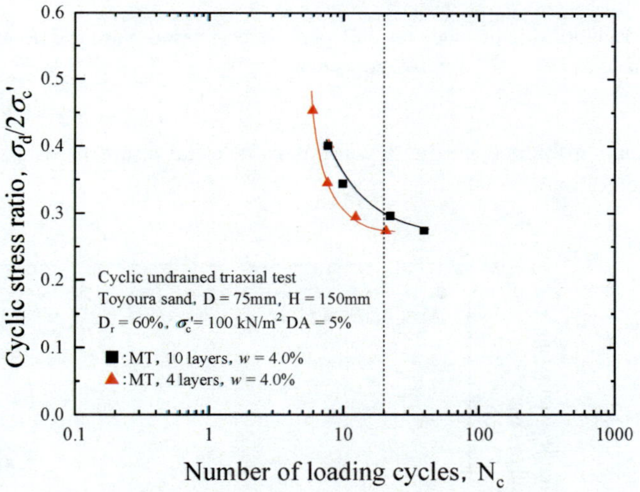

Fig. 5. Relationship between CSR and number of loading cycles for Toyoura sand

As shown in Fig. 5, in the case of Toyoura sand, the cyclic stress ratio, CSR ($N_c = 20$) for specimens with 10 layers is higher than that of specimens with 4 layers. However, the gradient of liquefaction curve is gentle for specimens with 10 layers but steep for specimens with 4 layers.

In general, liquefaction resistance of dense specimen will rise steeply, when the grain size distribution, specimen preparation method, etc. are the same conditions. The compaction thickness of specimen with 4 layers is relatively high, and the compaction energy may not reach the bottom edge in each layer, locally densely packed areas and locally loosely packed areas are assumed to form. Consequently, in the loosely packed areas, the excess pore water pressure and axial deformation may locally increase, and its area may become a weak one. In densely packed areas, the behavior might be the opposite results. In this study, it can be pointed out that the difference in density has a

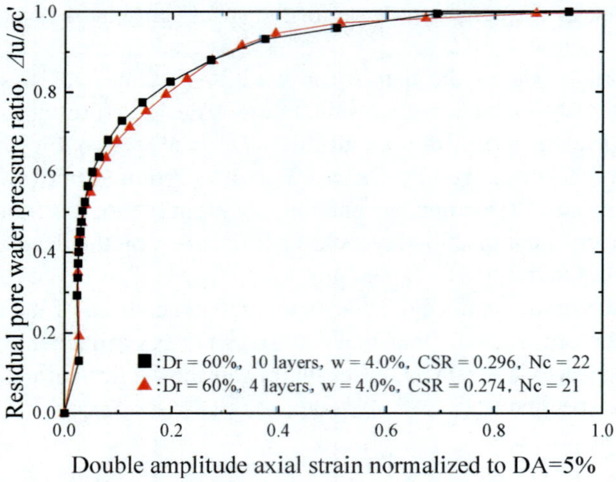

Fig. 6. Relationship between residual pore water pressure ratio and double amplitude axial strain normalized to $DA = 5\%$ for Toyoura sand

significant influence on the development of axial strain for the specimen with a thick compacted layer.

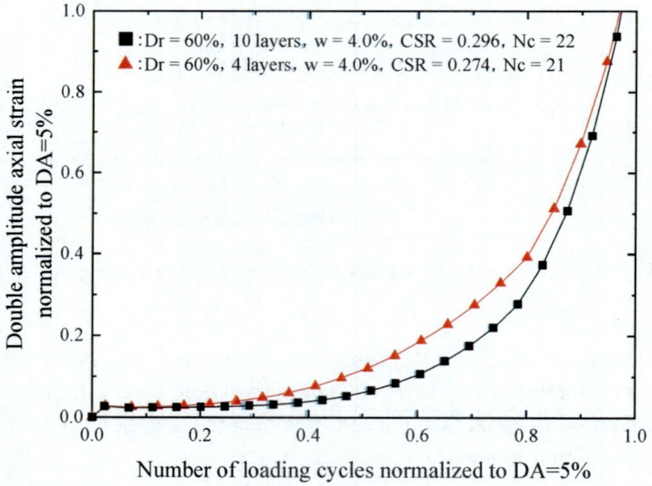

Fig. 7. Relationship between double amplitude axial strain normalized to $DA = 5\%$ and N_c required until $DA = 5\%$, N_c/N_c at $DA = 5\%$ for Toyoura sand

4.2 Liquefaction Properties of Well-Graded Sand

4.2.1 Liquefaction Resistance

Figure 8 shows the relationships between CSR and N_c required to *DA* of 5% for well-graded sand. It is confirmed from the figure that there is an apparent difference in liquefaction resistance comparing the three different degree of compaction, D_c.

In the consideration of the results of the specimens with 4 layers, the decrease rate at $N_c = 20$ based on $D_c = 95\%$, (CSR $_{Dc=95\%}$ - CSR)/ (CSR $_{Dc=95\%}$) was 41% for $D_c =$ 90% and was 55% for $D_c = 85\%$, respectively. This means that even if $D_c \geq 90\%$, only lowering D_c from 95% to 90%, the CSR ($N_c = 20$) decreases by 40%.

As with the Toyoura sand, it was found for well-graded sand to differ the liquefaction resistance according to differences in compaction thickness. However, the results showed a trend in which CSR ($N_c = 20$) decreases in the order of 4 layer, 10 layers, 1 layers. Therefore, the relationship between CSR ($N_c = 20$) and compaction thickness is different from Toyoura sand. In addition, the decrease in liquefaction resistance when using 4 layers as a standard is more evident in specimens with 1 layer than in specimens with 10 layers.

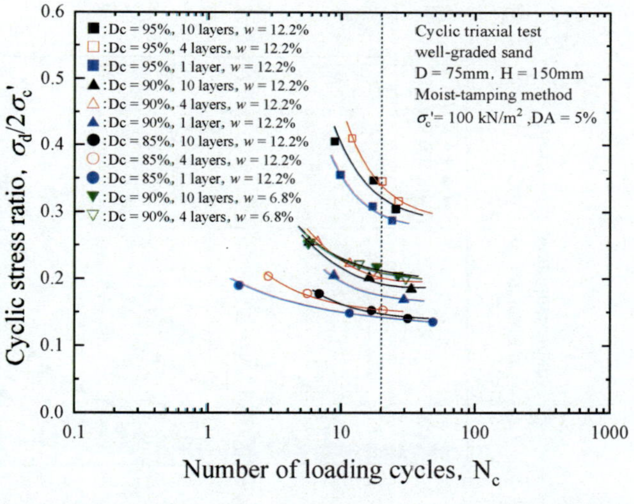

Fig. 8. Relationship between CSR and number of loading cycles for well-graded sand

Comparing specimens with 4 layers and 10 layers, a difference in liquefaction resistance can be confirmed for $D_c = 95\%$; however, there is almost no difference in liquefaction resistance for $D_c = 85\%$. This seems to be indicated that the influence in low density becomes dominant among the various factors affecting the mechanical properties for loose specimen. In the results of Toyoura sand, the CSR ($N_c = 20$) values are higher for specimens with 10 layers than with 4 layers. However, in the results of well-graded sand, slightly higher CSR ($N_c = 20$) are obtained for the specimens with 4 layers than with 10 layers, especially for $D_c = 95\%$. Therefore, even for the specimens prepared by the same compaction thickness, it was found that the difference in the results depending on the test materials appeared.

In the case of $D_c = 90\%$, two different initial moisture contents were set. For compacted soils under the same energy, it has been known that the initial moisture contents affect various mechanical properties (Miura and Toki 1984; Kawajiri et al. 2011; Yokohama et al. 2014).

In general, the largest strength of the compacted soils is considered to be obtained in moisture content slightly drier than that with w_{opt}, and the lowest value of coefficient of permeability is considered to be obtained in moisture content slightly wetter than that in w_{opt}. In this study, the CSR ($N_c = 20$) for specimens in w_n is slightly higher than that in w_{opt}, however the gradients of liquefaction curves were slightly gentler with specimen in w_n than that in w_{opt}.

4.2.2 Development of Axial Strain

Figure 9 shows the developments of *DA* until *DA* = 5% for representative test cases, where the data of N_c (*DA* = 5%) = 30 is adopted. Although the cases of 4 layers for D_c = 90% and 85% were omitted, because there was no corresponding data. The figure shows that the developments of axial strain until *DA* = 5% differ clearly due to the difference in specimen preparation conditions.

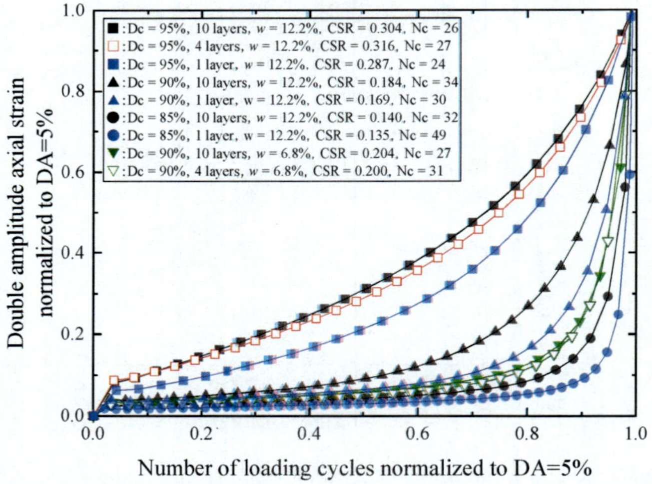

Fig. 9. Relationship between double amplitude axial strain normalized to *DA* = 5% and N_c required until *DA* = 5%, N_c/N_c at *DA* = 5% for well-graded sand

In this figure, it is evident that the change in axial strain until *DA* = 5% is more gentle for the higher the D_c of the specimen. It can be also seen that the curves in axial strain for $D_c = 95\%$ is more gradual with an increase of compaction layers; however the difference between 10 and 4 layers is small. On the other hand, the deformation behavior for specimens with 1 layer are similar to that of loose specimens, and the difference in the development of axial strain is significant compared with other compacted layers of 4 and 10. A similar tendency are also obtained in other D_c.

In the comparison of CSR ($N_c = 20$) in Fig. 8, the specimen with 4 layers indicates the highest values. In the results of the same specimen in Fig. 9, the case with 10 layers shows the most gradual development of *DA*. Therefore, a difference in the results for strength properties and deformation properties is evident.

In the case of $D_c = 90\%$, it is also obvious that the results differ depending on the difference in the initial moisture content during compaction. Comparing with specimens prepared with w_{opt} and specimens prepared with w_n, the inflection point of the curve for specimens in w_n is clearly appeared than that in w_{opt}, and the deformation behavior is similar to that of loose specimens. However, it is confirmed in Fig. 8 that specimens with w_n have slightly higher CSR ($N_c = 20$) than specimens with w_{opt}.

Therefore, the change in the initial moisture content had no significant effect on liquefaction resistance, but on deformation properties. The influence of differences in initial moisture contents on the development of axial strain of specimens has also been elucidated in research on sandy-silt soils, and vulnerability of specimens prepared in drier moisture contents than w_{opt} has been demonstrated (Matsumura et al. 2012).

4.2.3 Accumulation of Residual Pore Water Pressure

Figure 10(a) shows the relationship between the accumulation of $\Delta u/\sigma_c'$ during cyclic loadings and the development of *DA* until $DA = 5\%$. These data in this figure are from the same specimen as in Fig. 9. To observe the behavior of deformation and residual pore water pressure at the initial loading, Fig. 10(b) depicts the test results up to DA/DA $(5\%) = 0.10$.

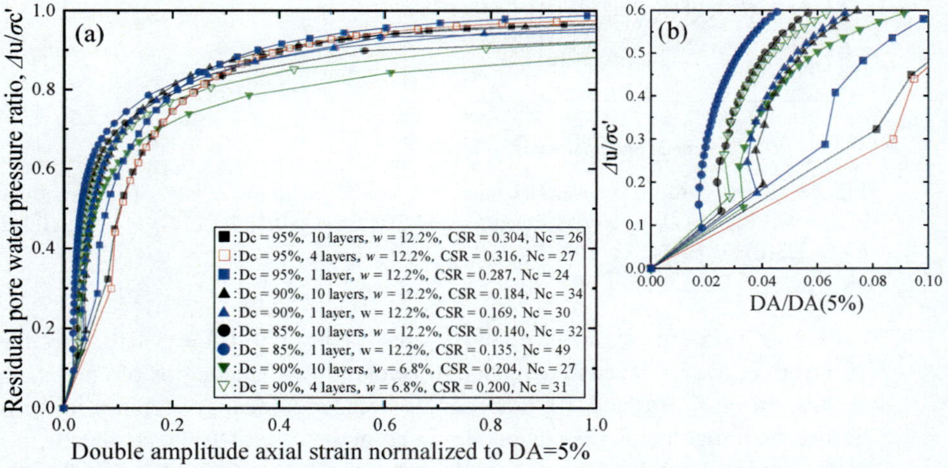

Fig. 10. Relationship between residual pore water pressure ratio and double amplitude axial strain normalized to $DA = 5\%$ for well-graded sand: (a) DA/DA $(5\%) \leq 1.0$, (b) DA/DA $(5\%) \leq 0.10$

Focusing on the difference in D_c, for any compaction thickness, the increase of $\Delta u/\sigma_c'$ is remarkable in the lower value of D_c relative to the development of axial strain. Their curves intersect around DA/DA $(5\%) = 0.2$, $\Delta u/\sigma_c'$ is closer to1.0 in the higher value of D_c.

In addition, the difference in accumulation of $\Delta u/\sigma_c'$ between 10 layers and 4 layers in the case of $D_c = 95\%$ is slight. However, comparing 10 layers and 1 layer, accumulation of $\Delta u/\sigma_c'$ of 1 layer is faster than 10 layers at the initial loading (see Fig. 10(b)).

In comparison to the 4 cases prepared with $D_c = 90\%$, the accumulation of $\Delta u/\sigma_c'$ under the same *DA*/*DA* (5%) is faster in the order of 4 layers (w_n), 10 layers (w_n), 1 layer (w_{opt}) and 10 layers (w_{opt}). That is that the $\Delta u/\sigma_c'$ at the same *DA*/*DA* (5%) increases with increasing the compaction thickness for the same moisture content. Similarly, the $\Delta u/\sigma_c'$ at the same *DA*/*DA* (5%) increases with decreasing moisture content for the same compaction thickness.

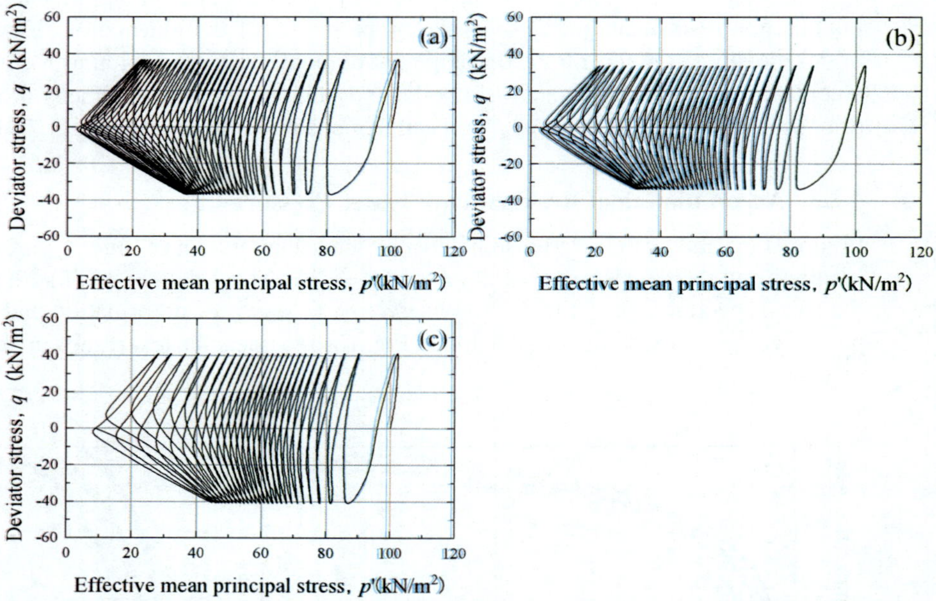

Fig. 11. Stress path for well-graded sand (*DA* <= 5%): (a) $D_c = 90\%$, 10 layers, $w = 12.2\%$, CSR = 0.184), (b) $D_c = 90\%$, 1 layer, $w = 12.2\%$, CSR = 0.169), (c) $D_c = 90\%$, 10 layers, $w = 6.8\%$, CSR = 0.204)

Figure 11(a)–(c) ,show the effective stress paths up to $DA = 5\%$ for specimens with 10 layers in w_{opt}, 1 layer in w_{opt} and 10 layers in w_n. The degree of compaction is $D_c = 90\%$. In Fig. 11(a) and (b), it appears to be no significant difference in the behavior during the initial loading behavior. However, in the case of 10 layers, cyclic mobility is maintained even after $\Delta u/\sigma_c'$ rises with a progresses of cyclic loading. On the other hand, the specimen with 1 layer did not indicate the resistance to deformation as specimens with 10 layers.

The effect of the difference in initial moisture content on the stress path is also investigated in Fig. 11(a) and (c). For comparison with the specimen with w_{opt} (Fig. 11(a)), the specimen with w_n (Fig. 11(c)) reaches suddenly 5% while the effective mean principal stress is decreasing. In Fig. 8, although the CSR of the specimen with w_n showed higher than that of the specimen with w_{opt}, it was found that the specimen with w_n is low toughness according to the results of Figs. 9, 10, 11.

These results show that the trend in the increase of $\Delta u/\sigma_c'$ during the initial loading of cyclic loadings for specimens prepared with 1 layer is similar to those of specimens prepared with the lower D_c and the drier moisture content. However, the $\Delta u/\sigma_c'$ at the final stage for the specimens prepared with 1 layer is more increased than those with the specimens in 10 layers. This tendency differs from the results from specimens with relatively brittle properties, like those prepared with a low D_c and drier moisture content.

Based on the above results, it can be seen that the effect of difference in the compaction thickness on the liquefaction properties for well-graded sand is significant and is different from that of a clean sand such as Toyoura sand.

5 Conclusions

In order to elucidate the effects of specimen preparation methods on the liquefaction properties of two types of sandy soils, a series of cyclic undrained triaxial tests was performed. The conclusions obtained are as follows:

1) Liquefaction resistance is influenced by the compaction thickness, and its tendency is changed depending on the grain size distribution and D_c. In the case of Toyoura sand, the number of layers tends to affect the liquefaction properties. In the case of well-graded sand, the liquefaction properties of specimens with 4 layers and with 10 layers was similar, but specimens with 1 layer showed a different tendency.

2) In the case of well-graded sands, the difference in initial moisture content had an influence on their deformation properties during cyclic loadings. In particular, the deformation behavior of specimens compacted with lower moisture content was similar to that of loose specimen.

The above results show that even with a specimen with the height of 150 mm, the specimen exhibits non-uniformity depending on the specimen preparation conditions. It can affect the mechanical behavior. Therefore, when preparing a uniform specimen, it is desirable that the compaction thickness be as thin as possible.

Further consideration will be required in the future. As part of that, we plan to perform an image analysis using the specimens prepared under different compaction thickness.

References

Kawajiri, S., Kawaguchi, T., Shibuya, S., Takahashi, S.: Effects of moulding water content and compaction method on deformation and strength characteristics of compacted soil. J. Jpn. Soc. Civ. Eng. Ser. C. (Geosphere Eng.) **67**(4), 532–543 (2011). (in Japanese)

Matsumura, S., Miura, S., Yokohama, S.: Effects of compaction on cyclic undrained shear properties of sandy-silt soils and its evaluation. J. Jpn. Soc. Civ. Eng. Ser. C. (Geosphere Eng.) **68**(4), 597–609 (2012). (in Japanese)

Miura, S., Toki, S.: Anisotropy in mechanical properties and its simulation of sands sampled from natural deposits. Soils Found. **24**(3), 69–84 (1984)

Tatsuoka, F., Ochi, K., Fujii, S., Okamoto, M.: Cyclic undrained triaxial and torsional shear strength of sands for different sample preparation methods. Soils Found. **26**(3), 23–41 (1986)

Sasaki, T., Kawamura, S., Koseki, J.: Effects of compaction thickness on liquefaction properties of two kinds of sandy soils prepared by moist-tamping method. In: 8th Japan-China Geotechnical Symposium (2020) (in press)

Yokohama, S., Miura, S., Matsumura, S.: Change in the hydromechanical characteristics of embankment material due to compaction state conditions. Soils Found. **54**(4), 731–747 (2014)

The Research of Dynamics Property and Long-Term Settlement About Weathered Argillaceous Siltstone Under Base of Subway Tunnel

Yu-Feng Shi[1,2], Chao Yu[3], Xiang-Sheng Chen[1,2](✉), Xiu-Shao Zhao[1,2], and Bi-Tang Zhu[1,2]

[1] National Experimental Teaching Demonstration Center of Civil Engineering, East China Jiaotong University, Nanchang 330013, Jiangxi, China
[2] Jiangxi Key Laboratory of Infrastructure Safety Control in Geotechnical Engineering, Nanchang 330013, China
[3] Foshan Railway Investment and Construction Group Co., Ltd., Foshan 52800, China

Abstract. Subway tunnel of Nanchang may pass through one or several weathered argillaceous siltstone in the construction process. To reveal the influence of weathered argillaceous siltstone on the long-term settlement of subway tunnel, the dynamic triaxial tests of intensely and moderately weathered argillaceous siltstone were carried out. Based on these tests results, the evolution law of cumulative plastic dynamic strain was analyzed and a prediction model of cumulative plastic strain was established. Finally, an actual subway tunnel with weathered argillaceous siltstone was selected for long-term settlement prediction analysis. The prediction data shows that the sedimentation value of tunnel is approximately 20.87 mm under 100 years operation. Research results indicate that the dynamic strain of weathered argillaceous siltstone under cyclic vibration load is mainly affected by dynamic stress and static deviatoric stress. When the static deviatoric stress increased from 0 kPa to 200 kPa, the dynamic strain of intensely and moderately weathered argillaceous siltstone increased by 8.53 times and 5.97 times, respectively. Therefore, the intensely weathered argillaceous siltstone (IWAS) is more susceptible to static deviatoric stress than moderately weathered argillaceous siltstone (MWAS). In addition, increasing the confining pressure will accelerate the expansion of fissures in intensely weathered argillaceous siltstone, while the fissures in moderately weathered argillaceous siltstone will be inhibited, which show diametrically opposite results.

Keywords: Tunnel engineering · Weathered soft rock · Cyclic load · Dynamic deformation · Nonlinear fitting · Long-term settlement

1 Introduction

There are more than 50% of basement of Nanchang subway tunnel located in argillaceous siltstone with different weathering degrees (mainly moderately weathered rock). This

Y. Liu et al. (Eds.): GeoChina 2021, SUCI, pp. 31–43, 2021.
https://doi.org/10.1007/978-3-030-80316-2_4

type of tunnel basement is likely to suffer structural damage under the long-term cyclic loading of train because of its characteristics of easy softening, uneven distribution, and fragility (Fig. 1). Therefore, it is necessary to invest great repair costs to ensure durability and safety of this tunnel basement. At present, many scholars are carrying out research on the long-term settlement of foundation caused by subway operation, especially subway tunnels in soft soil and silt soil [1]. However, the long-term settlement of subway tunnel in different weathered soft rocks have less concerned [2], which leads to a series of engineering problems for operating subway in such strata, such as differential settlement.

Fig. 1. Engineering problems of subway tunnel in weathered argillaceous siltstone

Due to the geological complexity, the cumulative settlement prediction of subway tunnel base under the action of cyclic dynamic loads is relatively complicated. The research in this area is currently focused on soft soli subgrades, with little attention paid to soft rock subgrades. Seed and Chan et al. [3] studied the dynamic characteristics of clay under cyclic loading by laboratorial experiments. Larew [4] found the existence of critical cyclic stress ratio when studying the dynamic strain of soft clay. Kazuya [5] found that irreversible plastic strain of soft soil under the action of cyclic load. Liu et al. [6] studied the trend of cumulative plastic deformation of gravel soil by dynamic triaxial test. Yang et al. [7] studied the cumulative plastic strain of muddy soft clay in Ningbo under different confining pressure and dynamic stress conditions. Wang et al. [8] studied the strength problem of silt clay in Yantai under the effects of different consolidation static and dynamic stresses. Ding [9] and Zhao et al. [10]. studied the dynamic characteristics of sandy mudstone under cyclic load and predicted the long-term settlement. Zhao [11] carried out mechanical test of fully weathered argillaceous siltstone, test results indicated that the weathered argillaceous siltstone has low strength and cannot be used as a roadbed filler for high-speed railways. Zhang et al. [12] carried out laboratory tests on argillaceous siltstones with different degrees of saturation. The test results show that water has little effect on the ultimate destruction morphology of argillaceous siltstones, but has a great influence on its microstructure. Chen et al. [13] studied the effect of cyclic frequency on the undrained performance of undisturbed marine clay based on cyclic triaxial test. Huang et al. [14] studied the long-term settlement of subway tunnel in saturated clay under the effects of vibration load. The results show that the final settlement of subway tunnels in saturated clay can reach 80 mm.

Since the subway tunnels in Nanchang are generally located in weathered argillaceous siltstone, which is different from the soft soil and sand foundations. Therefore, it is necessary to carry out laboratorial dynamic triaxial tests on different weathered argillaceous siltstones to reveal their dynamic deformation characteristics so as to accurately predict the long-term settlement of subway tunnels.

2 Dynamic Deformation Property of Weathered Argillaceous Siltstone

To investigate the dynamic characteristics of weathered argillaceous siltstone in different working conditions, the dynamic triaxial test of intensely and moderately weathered argillaceous siltstone was carried out with the effects of dynamic stress, static deviatoric stress, confining pressure and load frequency, and its deformation properties were also analyzed. The detailed working conditions and test parameters are shown in Table 1.

Table 1. Test parameters of moderately and intensely weathered argillaceous siltstone under dynamic triaxial test

Samples	Working condition	Dynamic stress amplitude (kPa)	Static deviatoric stress (kPa)	Confining pressure (kPa)	Vibration frequency
MWAS	1	55, 75, 100, 130, 170	100	500	1
	2	55	0, 50, 100, 150, 200	500	1
	3	55	100	350, 400, 450, 500, 550	1
	4	55	100	500	1, 1.5, 2, 5
IWAS	5	55, 75, 100, 130, 170	100	500	1
	6	55	0, 50, 100, 150, 200	350	1
	7	55	100	300, 350, 400, 450, 500	1
	8	55	100	350	1, 1.5, 2, 5

2.1 The Effects of Dynamic Stress Amplitude

In order to investigate the effects of dynamic stress amplitude on the strength of argillaceous siltstone, dynamic triaxial tests with dynamic stress amplitudes of 55 kPa, 75 kPa, 100 kPa, 130 kPa and 170 kPa were carried out on intensely and moderately weathered rock samples (working condition 1 and 5, Table 1), respectively. The confining pressure of intensely and moderately weathered argillaceous siltstone are 350 kPa and 500 kPa. The static deviatoric stress, loading time and frequency are 100 kPa, 5000 s and 1 Hz. The trend of axial cumulative plastic strain of intensely and moderately weathered argillaceous siltstone with the number of cyclic loading is shown in Fig. 2.

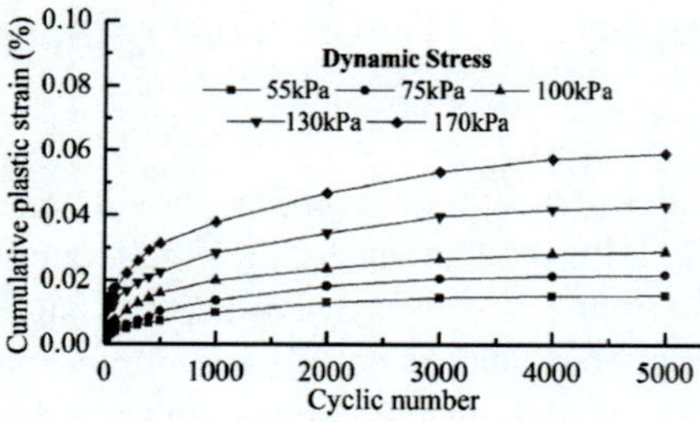

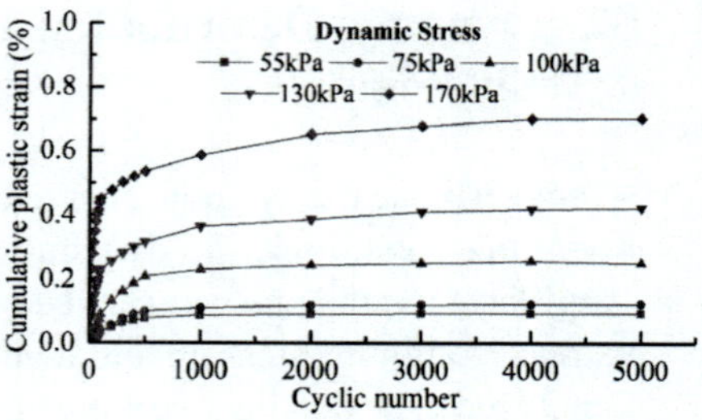

Fig. 2. Left: the cumulative plastic strain of moderately weathered argillaceous siltstone with the number of cyclic loading under different dynamic stress amplitudes. Right: the cumulative plastic strain of intensely weathered argillaceous siltstone with the number of cyclic loading under different dynamic stress amplitudes.

As can be seen from Fig. 2, the cumulative plastic strain of moderately weathered argillaceous siltstone will change with the dynamic stress level, and the number of cyclic loading when the cumulative plastic strain tends to be stable is 2000. Assuming that the cumulative plastic strain after 5000 cycles of loading is the total deformation, then when the dynamic stress amplitudes are 55 kPa, 75 kPa, 100 kPa, 130 kPa and 170 kPa, the corresponding stable cumulative strains (2000 cycles loading) account for approximately 84.51%, 83.11%, 81.66%, 80.42% and 79.19% of total strain (5000 cycles loading), respectively. However, the number of cyclic loading required when the cumulative plastic strain of the intensely weathered argillaceous siltstone (IWAS) stabilizes is far less than that of the moderately weathered argillaceous siltstone (MWAS), which is about 1000 cycles loading. When the dynamic stress amplitudes are 55 kPa, 75 kPa, 100 kPa, 130 kPa and 170 kPa, the corresponding stable cumulative strains (1000 cycles loading) of intensely weathered argillaceous siltstone account for approximately 91.98%, 90.10%, 88.14%, 86.02% and 83.43% of total strain (5000 cycles loading), respectively. Therefore, with the increase of dynamic stress amplitude, the growth trend of cumulative plastic strain of intensely and moderately weathered argillaceous siltstone is consistent. The difference is that the intensely weathered argillaceous siltstone reaches the stable plastic strain earlier than moderately weathered argillaceous siltstone.

2.2 The Effects of Static Deviatoric Stress

In order to investigate the effects of static deviatoric stress on the dynamic strain of argillaceous siltstone, dynamic triaxial tests with static deviatoric stress of 0 kPa and 200 kPa were carried out on intensely and moderately weathered argillaceous siltstone samples (working condition 2 and 6, Table 1). As shown in Fig. 3, with the increase of static deviatoric stress, the cumulative plastic strain of intensely and moderately weathered argillaceous siltstone will gradually increase and eventually stabilize. When the static deviatoric stress is 0 kPa and 200 kPa respectively, the final cumulative plastic strain of moderately weathered argillaceous siltstone is 0.0058% and 0.0349%, and the final cumulative plastic strain of intensely weathered argillaceous siltstone is 0.052% and 0.441%. It can be seen that under the same static deviatoric stress conditions, the cumulative plastic strain of intensely weathered argillaceous siltstone is approximately

10 times that of moderately weathered argillaceous siltstone, which means that intensely weathered argillaceous siltstone is more affected by the static deviatoric stress.

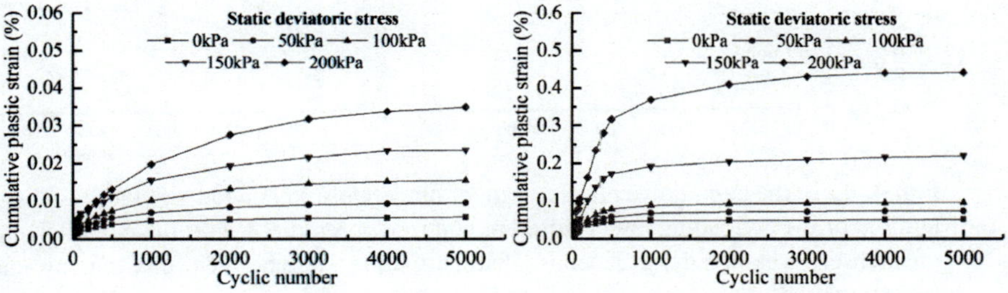

Fig. 3. Left: the cumulative plastic strain of moderately weathered argillaceous siltstone with the number of cyclic loading under different static deviatoric stress. Right: the cumulative plastic strain of intensely weathered argillaceous siltstone with the number of cyclic loading under different static deviatoric stress.

2.3 The Effects of Confining Pressure

Under the test conditions of dynamic stress amplitude is 55 kPa, static deviatoric stress is 100 kPa and loading frequency is 1 Hz, the dynamic triaxial tests of intensely and moderately weathered argillaceous siltstone with different confining pressure were carried out, test results are shown in Fig. 4. Compared with the dynamic stress amplitude and static deviatoric stress, confining pressure has little effect on cumulative plastic strain of weathered argillaceous siltstone. When the confining pressure is 350 kPa, 400 kPa, 450 kPa, 500 kPa, and 550 kPa respectively, the final cumulative plastic strain of moderately weathered argillaceous siltstone is 0.0169%, 0.0164%, 0.0160%, 0.0155% and 0.0147%. Therefore, the cumulative plastic strain of moderately weathered argillaceous siltstone will gradually decrease with the increase of confining pressure.

However, the effects of confining pressure on the cumulative plastic strain of intensely weathered argillaceous siltstone show an opposite results compared with that of moderately weathered argillaceous siltstone. Because test results indicate that as the confining pressure increases, the cumulative plastic strain of intensely weathered argillaceous siltstone will increase, which is mainly determined by the strength of rock itself.

2.4 The Effects of Load Frequency

Figure 5 is the curve of cumulative plastic strain of weathered argillaceous siltstone with the cyclic numer under different load frequency. When the load frequency is 1 Hz, 1.5 Hz, 2 Hz and 5 Hz respectively, the final cumulative plastic strain of moderately weathered argillaceous siltstone is 0.0155%, 0.0138%, 0.0104% and 0.0071%, and the final cumulative plastic strain of intensely weathered argillaceous siltstone is 0.0960%, 0.0901%, 0.0840% and 0.0397%. It can be seen that as the load frequency increases, the cumulative plastic strain of weathered rock will gradually decrease, which have a negative correlation. The test results are consistent with the research of scholar Ding [15].

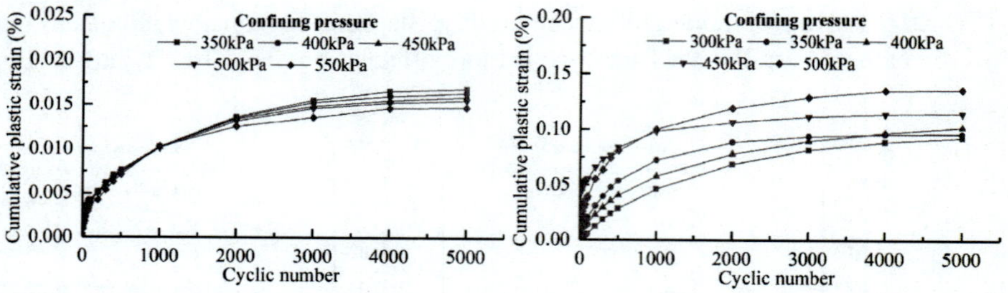

Fig. 4. Left: the cumulative plastic strain of moderately weathered argillaceous siltstone with the number of cyclic loading under different confining pressure. Right: the cumulative plastic strain of intensely weathered argillaceous siltstone with the number of cyclic loading under different confining pressure.

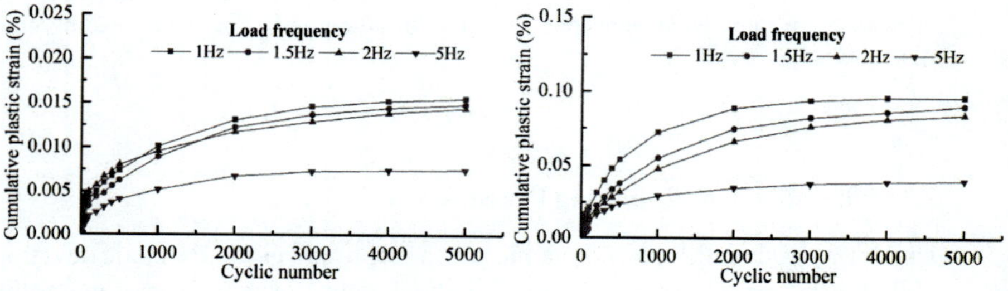

Fig. 5. Left: the cumulative plastic strain of moderately weathered argillaceous siltstone with the number of cyclic loading under different load frequency. Right: the cumulative plastic strain of intensely weathered argillaceous siltstone with the number of cyclic loading under different load frequency.

3 The Cumulative Plastic Strain Model

Based on the empirical formulas and the above test results of dynamic triaxial, the cumulative plastic strain model of weathered argillaceous siltstone can be established [15]. This model comprehensively considers the effects of dynamic stress, static stress and the number of cyclic loading on rock plastic strain.

$$\varepsilon_p = a(k\eta_s + \eta_d)^m N^b \tag{1}$$

where ε_p and N are the cumulative plastic strain and the number of cyclic loading, η_d and η_s are the dynamic stress ratio and static stress ration, a, k, m and b are the fitting parameters.

These fitting parameters (a, k, m and b) can be obtained by using mathematical analysis software. The fitting results are shown in Fig. 6 and Fig. 7, the goodness of fit (R^2) is above 0.92. According the fitting results, the cumulative plastic strain model of intensely and moderately weathered argillaceous siltstone can be obtained.

The cumulative plastic strain model of moderately weathered argillaceous siltstone:

$$\varepsilon_p = 0.1956(0.47\eta_s + \eta_d)^{1.773} N^{0.3306} \tag{2}$$

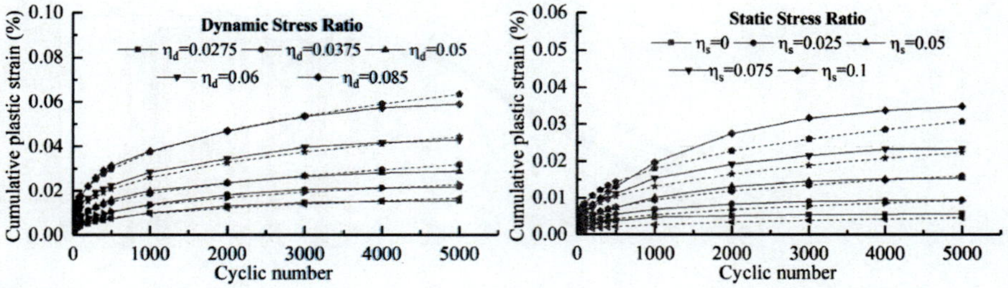

Fig. 6. Left: the test and fitted curve between cumulative plastic strain of moderately weathered argillaceous siltstone and cyclic number under different dynamic stress ratio. Right: the test and fitted curve between cumulative plastic strain of moderately weathered argillaceous siltstone and cyclic number under different static stress ratio.

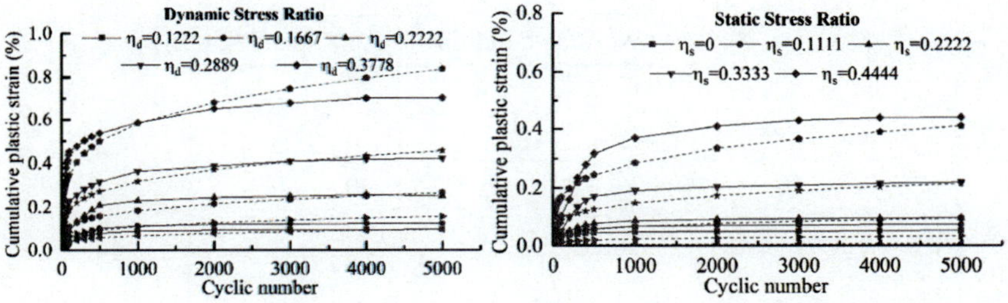

Fig. 7. Left: the test and fitted curve between cumulative plastic strain of intensely weathered argillaceous siltstone and cyclic number under different dynamic stress ratio. Right: the test and fitted curve between cumulative plastic strain of intensely weathered argillaceous siltstone and cyclic number under different static stress ratio.

The cumulative plastic strain model of intensely weathered argillaceous siltstone:

$$\varepsilon_p = 0.9919(0.6843\eta_s + \eta_d)^{3.29} N^{0.2255} \quad (3)$$

4 Dynamic Response Analysis of Weathered Argillaceous Siltstone

By using Midas GTS/NX to establish a three-dimensional numerical model of subway tunnel located in weathered argillaceous siltstone, the dynamic response characteristics of weathered argillaceous siltstone at the bottom of the tunnel under the train load were studied. The train load is mainly derived from the interaction between the wheels and rails during the travel process. Figure 8 shows the position of train load and time-history curve, this paper only considers the vertical load. According to the Nanchang rail transit vehicle model, the starting and ending point of train load are selected, and different axial spacing and axial load are also defined. The running conditions of this train are shown in Table 2.

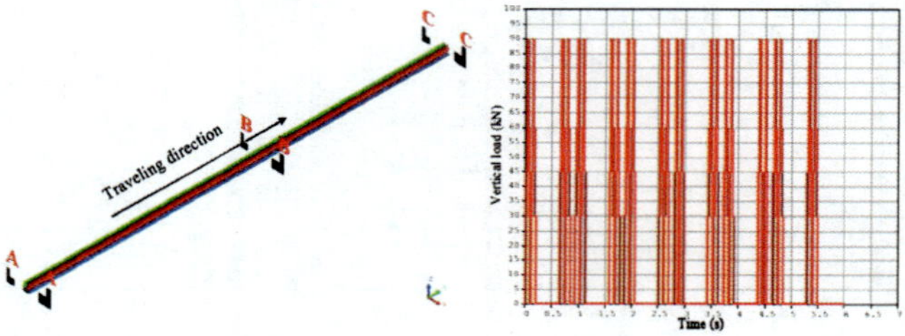

Fig. 8. Left: the traveling direction and train load position. Right: the time-history curve of train load in cross section A-A.

Table 2. Working conditions and train speed for dynamic response analysis of weathered argillaceous siltstone

Working conditions	Train speed (km/h)
1	50
2	70
3	90
4	120

The distribution of surrounding formations in this subway tunnel model is shown in Fig. 9, there are nine formations. The depth of this tunnel is about 22 m, the thickness of segment is 25 cm. In the subsequent analysis, four positions 0.5 m, 1 m, 2 m and 4 m from the outside of tunnel lining were selected as analysis points. The train passing time is 14.4 s, 10.28 s, 8 s and 6 s respectively.

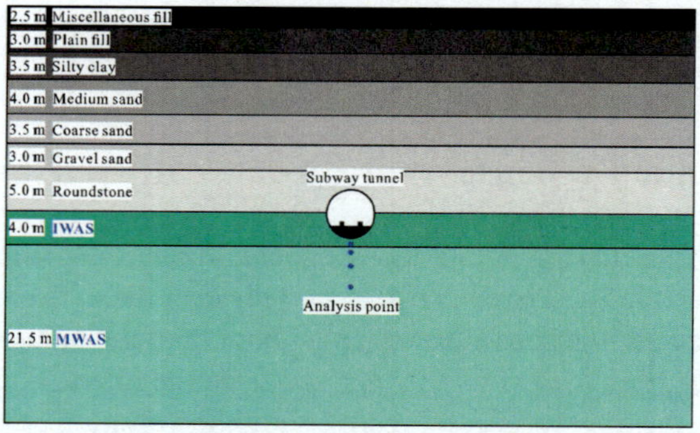

Fig. 9. The distribution and thickness of surrounding formations in the subway tunnel model

4.1 Acceleration Response Characteristics

The numerical analysis results show that with the increase of the train speed, the peak acceleration of the tunnel base will gradually increase (Fig. 10). The detailed calculation values of peak acceleration are shown in Table 3. Taking point A as an example, when the train speed is 50 km/s, 70 km/s, 90 km/s and 120 km/s, the peak acceleration of the tunnel base is 0.294 m/s^2, 0.346 m/s^2, 0.404 m/s^2 and 0.429 m/s^2, respectively. In addition, the peak acceleration of the tunnel base will gradually decrease with the increase of depth. When the train speed is 70 km/s, the peak acceleration of analysis point A to D will be reduced from 0.346% to 0.085%, a reduction of 75.4%.

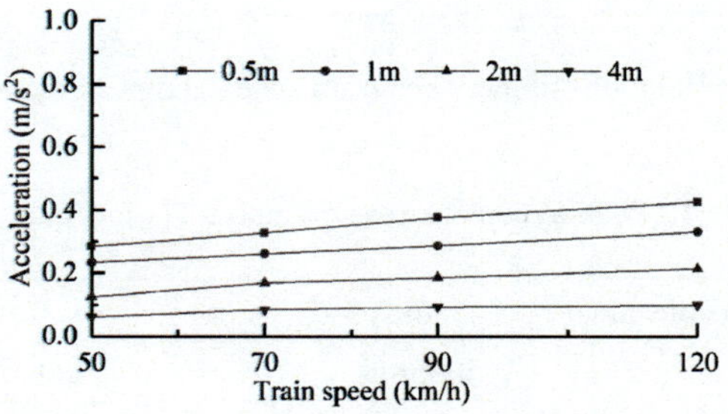

Fig. 10. Relationship between train speed and peak acceleration of tunnel base.

Table 3. Peak acceleration of analysis points under different train speed.

Analysis points of tunnel base	Train speed			
	50 km/h	70 km/h	90 km/h	120 km/h
	Acceleration (m/s^2)			
Point A (0.5 m outside the lining)	0.2938	0.3455	0.4039	0.4285
Point B (1.0 m outside the lining)	0.2441	0.2796	0.2953	0.3432
Point C (2.0 m outside the lining)	0.1333	0.1345	0.1361	0.1370
Point D (4.0 m outside the lining)	0.0693	0.0847	0.0933	0.0943

4.2 Dynamic Stress Response Characteristics

The relationships between peak dynamic stress value of tunnel base and train speed are shown in Fig. 11. As the train speed increases, the dynamic stress peak value will gradually increase. The detailed calculation values are shown in Table 4. Also taking point A as an example, when the train speed is 50 km/s, 70 km/s, 90 km/s and 120 km/s, the peak dynamic stress value is 21.80 kPa, 22.20 kPa, 24.30 kPa and 30.90 kPa, respectively. In addition, the peak dynamic stress value of tunnel base will gradually decrease with the increase of depth. When the train speed is 120 km/s, the peak dynamic stress value of analysis point A (0.5 m) and D (4.0 m) is 30.90 kPa and 18.30 kPa, the value of point A is about 1.69 times that of point D.

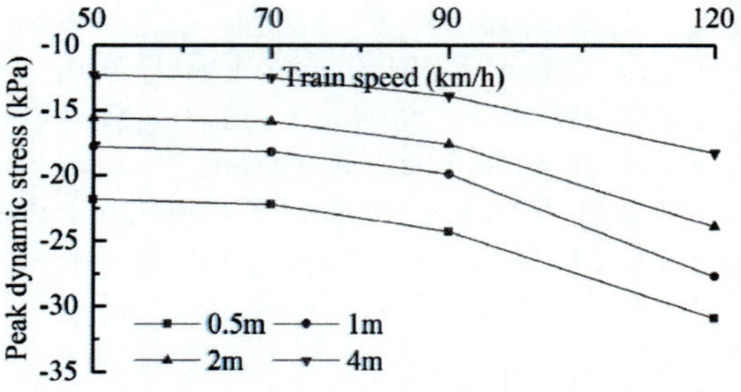

Fig. 11. Relationship between train speed and peak dynamic stress of tunnel base.

Table 4. Peak dynamic stress of analysis points under different train speed.

Analysis points of tunnel base	Train speed			
	50 km/h	70 km/h	90 km/h	120 km/h
	Dynamic stress (MPa)			
Point A (0.5 m outside the lining)	−0.4842	−0.4846	−0.4867	−0.4933
Point B (1.0 m outside the lining)	−0.5123	−0.5127	−0.5144	−0.5222
Point C (2.0 m outside the lining)	−0.5443	−0.5446	−0.5463	−0.5526
Point D (4.0 m outside the lining)	−0.6955	−0.6957	−0.6971	−0.7015

5 Longtime Cumulative Deformation Analysis of Weathered Argillaceous Siltstone

The cumulative plastic strain equations of (2) and (3) fitted by dynamic triaxial test data were used to analyze the longtime cumulative deformation of weathered argillaceous siltstone under the subway tunnel. Therefore, the vertical settlement deformation of tunnel base can be obtained by substituting the working conditions and calculation parameters into the cumulative plastic strain model. The working conditions and calculation parameters are shown in Table 5. The influence depth of train load is taken as 4 m below the tunnel base.

Analysis results show that with the increase of vibration frequency, the vertical settlement of the subway tunnel base will gradually increase and eventually tend to be stable (Fig. 12). When the train speed is 50 km/h, the sedimentation value of subway tunnel base within 20 years is about 13.12 mm, which accounts for 62.89% of the total sedimentation value in 100 years of operation. Therefore, the sedimentation value of subway tunnel base will gradually increase with the operation time. The growth trend shows rapid growth in the early period, and it tends to be flat in the later period.

Table 5. Calculation parameters of cumulative plastic strain model for predicting the long-term settlement

Formations	Thickness (m)	50 km/h		70 km/h		90 km/h		120 km/h	
		η_d	η_s	η_d	η_s	η_d	η_s	η_d	η_s
IWAS	0.5	0.0291	0.6165	0.0296	0.6165	0.0304	0.6165	0.0324	0.6165
IWAS	0.5	0.0237	0.6593	0.0243	0.6593	0.0251	0.6593	0.0265	0.6593
MWAS	1.0	0.0078	0.2644	0.0080	0.2644	0.0082	0.2644	0.0088	0.2644
MWAS	2.0	0.0062	0.3416	0.0063	0.3416	0.0064	0.3416	0.0070	0.3416

When the train speed is 50 km/s, 70 km/s, 90 km/s and 120 km/s, the long-time settlement of tunnel base is 20.85 mm, 20.87 mm, 20.89 mm and 20.96 mm. It can be seen that the train speed has little effect on the longtime settlement deformation of tunnel base. The reason is mainly related to the soil bearing capacity of tunnel base.

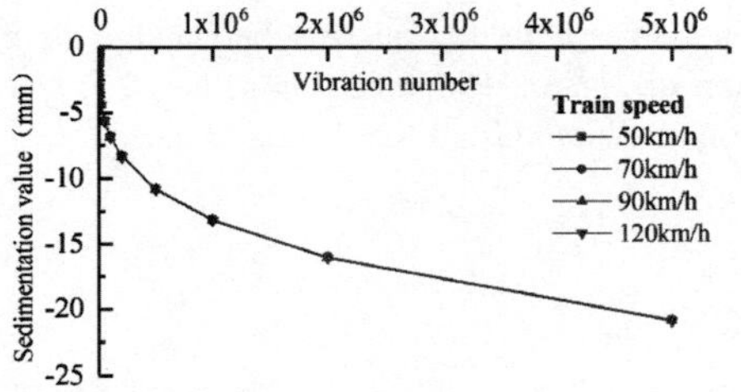

Fig. 12. Relationship between vibration number and sedimentation value of tunnel base under different train speed.

6 Conclusions

(1) The factors influencing the deformation of weathered argillaceous siltstone include dynamic stress amplitude, static deviatoric stress, confining pressure and load frequency, among which the dynamic stress amplitude has the most significant effect on the deformation. With the increase of dynamic stress amplitude, the cumulative plastic strain of intensely and moderately weathered argillaceous siltstone will gradually increase, and the growth rate and amount of intensely weathered argillaceous siltstone are greater than that of moderately weathered argillaceous siltstone.

(2) Due to the different weathering degree and fissure development of intensely and moderately weathered argillaceous siltstone, the cumulative plastic strain of them show a completely opposite trend when affected by confining pressure. The cumulative plastic strain of moderately weathered argillaceous siltstone is negatively correlated with the confining pressure, while the cumulative plastic strain of intensely weathered argillaceous siltstone is positively correlated with the confining pressure.

(3) Under the test conditions of dynamic stress amplitude is 55 kPa, static deviatoric stress is 100 kPa and loading frequency is 1 Hz, the final cumulative plastic strain of intensely weathered argillaceous siltstone is about 5.7 times that of moderately weathered argillaceous siltstone. Therefore, Nanchang subway tunnel may suffer the differential settlement when it passes through one or several weathered argillaceous siltstone with different weathering degrees, which will lead to the damage of shield tunnel structure.
(4) The dynamic response index of tunnel base will increase with the increase of the train speed, such as acceleration response and dynamic stress response. Compare with dynamic stress response, the acceleration response is more obviously affected by train speed.
(5) Based on the dynamic triaxial test and the dynamic response analysis of weathered argillaceous siltstone, the numerical prediction results show that when the Nanchang subway train speed is 70 km/h, the sedimentation value of tunnel base within 100 yeas of operation is approximately 20.87 mm.

Acknowledgements. The authors wish to acknowledge the financial supports from the Natural Science Foundation of Jiangxi Province (Grant No. 20192BAB206043), the National Natural Science Foundation of China (Grant No. 51768020).

References

1. Yang, A., Ma, R., Shao, C.: Experimental study on dynamic characteristics of structural soft clay based on different initial static deflection stress. Highway Transp. Sci. Technol. **34**(10), 19–25 (2017)
2. Chen, X., Wang, Y., Zhou, T., et al.: Long-term dynamic stability analysis on argillaceous siltstone physical improved soil subgrade of high-speed railway. J. Central South Univ. **43**(09), 3619–3624 (2012)
3. Seed. H.B., Chan, C.K., Monismith, C.L.: Effects of repeated loading on the strength and deformation of compacted clay. In: Highway Research Board Proceedings, pp. 541–558 (1955)
4. Larew, H., Leonards, G.: A repeated load strength criterion. In: Proceedings of the Highway Research Board, pp. 529–556 (1962)
5. Kazuya, Y., Toyotoshi, Y., Kazutoshi, H.: Cyclic strength and deformation of normally consolidation clay. Soils Found. **22**(3), 77–91 (1982)
6. Liu, D., Yang, X., Wang, J., et al.: Study on influence factors of gravel soil accumulative deformation under cyclic loading. J. Railway Sci. Eng. **11**(04), 68–72 (2014)
7. Yang, B., Liu, B.: Analysis of long-term settlement of shield tunnel in soft soil area under cyclic loading of subway train. China Railway Sci. **37**(03), 61–67 (2016)
8. Wang, Y., Lei, J., Li, Q., et al.: Post-cyclic strength degradation behavior of soft clay under anisotropic consolidation and numerical simulation. Chin. J. Geotech. Eng. **39**(09), 1557–1564 (2017)
9. Ding, Z., Peng, L., Shi, C., et al.: Experimental study on dynamic deformation behaviors of water-rich sandy mudstone under cyclic loading. Chin. J. Geotech. Eng. **34**(03), 534–539 (2012)

10. Zhao, D., Cao, Q., Huang, J.: Dynamic deformation of soft rock under cyclic loading in subway tunnel basal. J. Railway Sci. Eng. **13**(11), 2204–2210 (2016)
11. Zhao, Y.: Test and analysis of dynamic character of chemical improved soil of argillaceous siltstone for subgrade of high-speed railway. J. Railway Eng. Soc. **29**(03), 22–25+106 (2012)
12. Zhang, A., Zhang, Y., Liu, X., et al.: Test study on water affected to physical mechanics performances of muddy siltstone. Coal Sci. Technol. **43**(08), 67–71 (2015)
13. Chen, Y., Ji, M., Huang, B.: Effect of cyclic loading frequency on undrained behaviors of undisturbed marine clay. China Ocean Eng. **18**(4), 643–651 (2004)
14. Huang, Q., Huang, H.-W., Ye, B., et al.: Dynamic response and long-term settlement of a metro tunnel in saturated clay due to moving train load. Soils Found. **57**(6), 1059–1075 (2017)
15. Ding, Z.: Dynamic properties of soft rock and the safety of base structure of high-speed railway tunnels. Central South University (2012)

Tunnel Reinforcement in Soft Grounds Using Umbrella Arch Method

Ali Morovatdar[1(✉)], Massoud Palassi[2], and Kambiz Behnia[2]

[1] Department of Civil Engineering, University of Texas At El Paso, 500 W. University Ave, El Paso, TX 79968, USA
amorovatdar@miners.utep.edu

[2] College of Engineering, Faculty of Civil Engineering, University of Tehran, 14155-6619 Tehran, Iran
{mpalas,ircbehnia}@ut.ac.ir

Abstract. The efficient use of underground spaces through constructed tunnels offers a practical solution for the recent increase in the mobility demands. However, in the case of tunneling at shallow depths that the ground consists of weak soils, it deems necessary to deploy reinforcement techniques in order to maintain the stability and safety of the construction process. In this study, the influence of the Umbrella Arch Method (UAM) on controlling the tunneling-induced settlements under severe geotechnical conditions was assessed. To accomplish this objective, the tunnel excavation sequences, as well as the UAM deployment, were simulated through a series of finite element numerical modeling, considering the in-situ conditions associated with a case study. Post-processing of the numerical modeling results indicated that using the UAM resulted in a substantial reduction in the corresponding tunnel crown and ground surface settlements. Additionally, through a parametric analysis, design parameters of UAM pipes, in terms of geometrical and installation- related properties, were evaluated to provide relevant information on the optimized values for each individual parameter. The associated results showed that the optimum value of the pipe length is approximately 1.5 D_{tunnel} (the equivalent tunnel diameter). Additionally, for the range of design parameters evaluated in this study, the best performance of the UAM was achieved when the pipes diameter, length, installation angle, and installation distance were in the neighborhood of 12 cm, 12 m, 6°, and 30 cm, respectively.

Keywords: Umbrella Arch Method (UAM) · Tunnel reinforcement · Tunnel Stability · Forepoling pipes · Finite element analysis · Numerical simulation

1 Introduction

Due to the rapid population growth in the world in last few decades, the need for underground excavation and tunneling has been evolving (Zamanian et al. 2020). Roadway and railway tunnels have offered one of the most effective solution alternatives for transportation problems. Underground excavation needs specific considerations in different stages from design to operation and as a result, different excavation methods and design

Y. Liu et al. (Eds.): GeoChina 2021, SUCI, pp. 44–56, 2021.
https://doi.org/10.1007/978-3-030-80316-2_5

procedures have been introduced to take into account the stability and safety issues, for both excavations and adjacent infrastructures (Estahbanati et al. 2020). In order to find optimal excavation method and design variables satisfying the tunnel stability, design practitioners should properly evaluate the tunnel and ground settlements considering various design strategies to mitigate and control the excavation-induced settlements in actual field conditions. Otherwise, success in the construction of the tunneling project will be seriously jeopardized. Particularly, in the case of tunneling at shallow depths that the ground consists of soft soils (Salimi et al. 2020a, b) or weak rocks, safe construction of tunnel without causing any damages is of greatest importance. Therefore, one of the main challenges for engineers to construct tunnels in such critical regions is to limit the settlements to acceptable ranges. This is extremely important form both the tunnel stability and the safety viewpoints.

Construction of the roadway tunnels using the New Austrian Tunneling Method (NATM), in combination with auxiliary supporting techniques such as Umbrella Arch Method (UAM), jet grouting (Farhangi et al. 2020), and mechanical pre-cutting are commonly used to overcome the significant difficulties faced while tunneling in soft and weak ground. Specifically, UAM has been used to a great extent, due to its potential pre-reinforcement capabilities (Song et al. 2013; Schumacher and Kim 2013). In this approach, prior to tunnel excavation, a series of forepoling pipes are installed along the tunnel circumference in the crown. Subsequently, by injecting the grout through the pipes, the stiffened soil (between the pipes) coupled with the forepoling pipes create an umbrella-shaped arch above the tunnel. This arrangement significantly improves the stiffness properties of the impacted soil and maintains stability of the tunnel (Morovatdar et al. 2020b).

Several studies have been conducted to investigate the performance of the UAM in the tunnel construction procedure. Ranjbarnia et al. (2018) developed an analytical approach to evaluate the behavior of the UAM in deep tunnels. In the proposed approach, the displacement of the supported span of the tunnel was calculated using the convergence-confinement method. The results showed that the pipes diameter is a significant factor controlling the excavation-induced settlements in deep tunnels. Ocak (2008) studied the second stage excavation of Istanbul Metro, which was constructed using NATM in combination with umbrella arch method and resulted that using the UAM can effectively control the surface deformations, especially in clay-bearing formations. Aksoy and Onargan (2010) assessed the influence of implementing UAM and tunnel face bolts on ground surface settlements induced during the tunnel construction of the second phase of Izmir Metro, which was located in the densely populated district. The results of the numerical modeling indicated that the UAM and face bolt applications could considerably reduce the risk of buildings' failure by decreasing the ground settlements by 69%. Ocak and Selcuk (2017) deployed 2D numerical models to investigate the performance of the UAM in tunneling. It was revealed that although implementation of the UAM is time-consuming, it is a practical pre- supporting method that can significantly control the tunneling-induced displacements.

Additionally, considering the design parameters of the elements in the UAM reinforcing system, a number of studies have been carried out. Oke et al. (2014), using two-dimensional (2D) and three-dimensional (3D) models, investigated how numerical modeling can be implemented as an effective tool to evaluate the influential design parameters attributed to the use of UAM technique. The authors later developed a second-order equation for distributed load through a semi-analytical solution in which, it is assumed, the beam lays on the elastic foundation (Oke et al. 2016). Recently, Qian et al. (2019), using the Winkler-spring model, developed an analytical approach to simulate the behaviors of the UAM pipes. The authors found a relatively good agreement between the results obtained from the proposed model and the numerical modeling using FLAC 3D. Klotoé and Bourgeois (2019), conducting 3D finite element simulations, evaluated the influence of the UAM on the settlements induced by shallow tunneling. Through a parametric analysis, the authors found that the influence of using the UAM on the settlements remained modest for the range of parameters considered in the analysis.

Preceding studies in the literature have provided insights into the working mechanism and behavior of the tunnel reinforced by the umbrella arch technique. However, deploying non- uniform and inconsistent analysis approaches in some research studies for evaluation of the effect of using UAM in tunneling led to widespread results. In terms of previous numerical studies, one of the primary sources of inaccuracy could be attributed to the fact that some of these studies relied on several simplifying assumptions made in the analysis. For instance, in a number of numerical analyses, the elements consisting the primary support system of the tunnel, i.e., steel frame, shotcrete, and wire mesh, were simulated by an equivalent section representing the entire composite support system. Likewise, several studies have documented that the UAM elements, i.e., steel pipes, cement grout, and soilcrete, were simulated as one single element with equivalent material properties based on the weighted averages. The root of the problem attributed to these approaches is that the determination of the internal forces generated in the pipes would be extremely challenging. Consequently, making these simplifications, instead of modeling the aforementioned elements individually, simulate the circumstance that is vastly different from the actual field conditions, and hence, this can seriously jeopardize the accuracy of the analysis results.

In this study, through a comprehensive evaluation, the information on the tunneling procedure and UAM deployment from the actual field conditions coupled with lab-derived geotechnical properties of the existing soil were, in turn, incorporated into a series of FE simulation models using the ABAQUS program. A novel procedure was then devised to simulate all the UAM and primary support elements separately, with consideration of the appropriate interactions. Subsequently, the tunneling-induced settlements were calculated for further comparative analysis in this research effort. The second stage of the analysis in this study was to quantitatively assess the main impacts of pipe characteristics on the tunnel stability. Hence, forensic analyses were performed to extensively investigate the optimum values of different design parameters of the UAM pipes.

2 Methodology

The performance of the umbrella arch method in terms of stability of the tunneling practice was evaluated utilizing finite element analysis for a case study in Ghazvin-Rasht tunnel project (Morovatdar et al. 2020a). The resultant remarkable improvement of the tunnel stability justified applying this method in specific conditions. Hence, the research team decided to extensively investigate the impact of various design variables, e.g. geometric and installation parameters, on capability of UAM approach in controlling critical settlements. The parameters consisted of installation angle and overlap length, diameter, length and transverse distance of the pipes embedded in the tunnel crown. To achieve the mentioned purpose, several numerical simulations were carried out deploying ABAQUS/CAE (Simulia 2007). In these simulations, the effect of different pipe parameters on the induced settlements were evaluated by changing variables' values, recording the responses and post-processing the results. Following the simulations and obtaining the results, the sensitivity of the critical settlements to alteration of different prameters was illuminated by a sensitivity analysis performed. In the following sections, the UAM method, excavation procedure and installation of the pipes will be described, followed by which, the details of the numerical simulation, parameters taken into account and the results of the sensitivity analysis will be provided.

2.1 Case Study

Ghazvin-Rasht tunnel in Northern District of IRAN was considered as a case study. This arch-shaped tunnel with a height and width of 8 m is categorized into railway tunnels (Morovatdar et al. 2020). Geometric characteristics of the Ghazvin-Rasht railway tunnel are tabulated in Table 1.

Table 1. Geometric properties of Ghazvin-Rasht railway tunnel.

Geometric characteristic	Quantity
Length	1800 m
Longitudinal slope	−1.14%
Maximum overburden height	20 m
Tunnel span	8 m
Tunnel height	8 m

3 Numerical Modeling

In this section, different aspects of the finite element analysis are defined. These aspects include model geometry, primary supporting system components, UAM method and the design variables.

3.1 Model Geometry

Proper characterization of the dimensions and geometry of the model is of utmost importance to mitigate the systematic errors associated with boundary effect problems

(Morovatdar et al. 2019, 2020c; Salimi and Ghazavi 2019). To find the proper valu the model dimensions, a sensitivity analysis was carried out comprising of increa the height, width and length from 20 to 80 m and estimating the vertical settleme the crown until a point after which the crown vertical displacement changes neglig At this point of optimization, the height, width, and length of the model was arc 50, 50, and 70, respectively. These dimensions were used for all of the models adop continuum rigid elements as shown in Fig. 1.

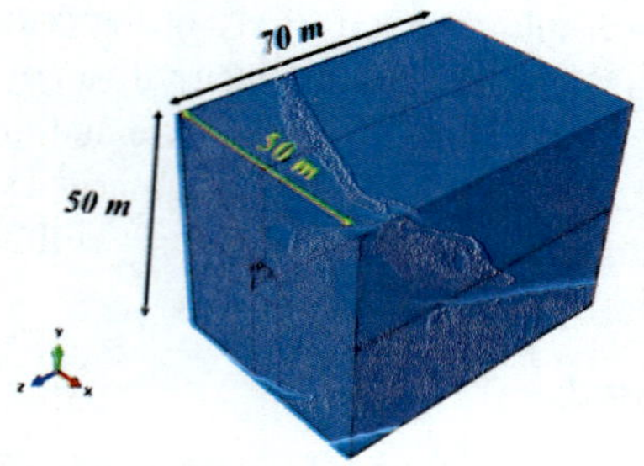

Fig. 1. Optimized model geometry in ABAQUS.

3.2 Mechanical Behaviors and Material Properties

The classic elastic-plastic, elastic and Mohr-Coulomb mechanical models were assigned to the steel frames and shotcrete, pipes, and soil, respectively. The linear elastic-perfectly plastic Mohr-Coulomb model is a fast and relatively accurate model; hence, it was adopted as the behavior model for soil and soilcrete (Beizaei et al. 2020; Beyzaei and Hosseininia 2019; Dehghani et al. 2020; Alimohammadi and Abu-Farsakh 2019; Farhangi and Karakouzian 2019; Satvati et al. 2020). Defined mechanical behavior models are summarized in Table 2.

Table 2. Mechanical behaviors of different elements.

Element	Soil	Soilcrete	Shotcrete	Steel Frame	Pipes
Mechanical behavior	Mohr-Coulomb	Mohr-Coulomb	Classic Elastic-Plastic	Classic Elastic-Plastic	Elastic

Additionally, a number of boreholes were drilled in the studied site to obtain the desirable geotechnical characteristics. The post-processed results from the laboratory experiments on the samples extracted from the boreholes, in terms of the geotechnical properties of the soil, are indicated in Table 3. Moreover, to characterize these properties associated with the jet grouted soilcrete at the vicinity of the tunnel, a few core samples were taken from the formed umbrella at the tunnel crown region. Table 3 also indicated the analyzed results attributed to the soilcrete zone. Table 4 also showed the material properties associated with the primary supporting system as well as the forepoling pipes (Morovatdar et al. 2020a).

Table 3. Geotechnical properties attributed to the soil and soilcrete.

Parameter	Material		Unit
	Soil	Soilcrete	
Density (ρ)	1900	2300	Kg/m^3
Young's modulus (E)	5	2600	MPa
Poisson's ratio (υ)	0.3	0.3	-
Cohesion (C)	40	4200	KPa
Internal friction angle (φ)	30	38	Degree

Table 4. Mechanical properties of the supporting system components.

Parameter	Density (Kg/m^3)	Young's modulus (GPa)	Poisson's ratio	28-day strength (MPa)
Shotcrete	2200	26	0.3	30
Steel frame	7800	200	0.25	-
Pipes	3200	210	0.35	-

3.3 Primary Supporting System Components

20 cm of shotcrete and steel frames of IPE18 were used as the primary supporting system. In the simulation of different cases, first one meter of the tunnel was excavated using the NATM (Wittke et al. 2006). Then the primary support system was installed in the crown and sides of the tunnel and this sequence was repeated until achieving the total length of the tunnel. The dimensions of the U-shaped frames was 8 × 8 m. The following steps describe the tunnel excavation sequences:

- Excavating one meter of the tunnel,
- Installing the associated primary support system.

After each excavation steps, wire elements were adopted to model steel sets having intervals of one meter. After the excavation completed, shotcrete was used to cover walls and invert of the tunnel as well as the face of the excavation.

3.4 Umbrella arch Method and Design Variables

In the modeling procedure, the UAM elements namely soilcrete, cement grout and the pipes were considered separately which are shown in Fig. 2. It should be noted that soilcrete refers to the soil around the pipes that is improved. In addition, the geotechnical characteristics and soil strength parameters adopted in the analysis were derived from laboratory experiments results of the samples taken from the drilled bore holes on site (Morovatdar et al. 2020a).

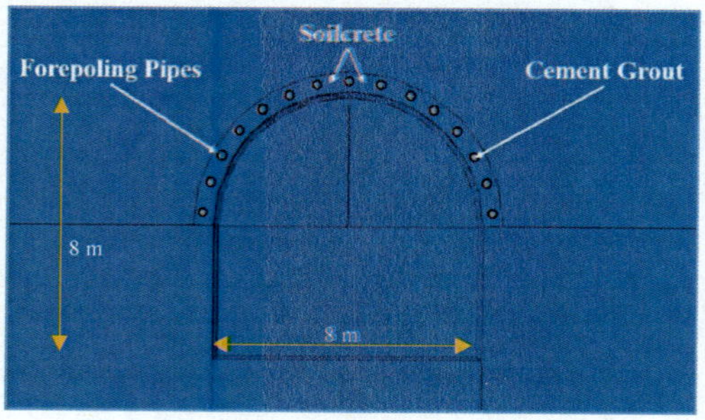

Fig. 2. Umbrella arch components.

As mentioned before, to investigate the impact of different parameters pertinent to forepoling pipes on the tunnel stability and crown settlements, diameter, length and installation distance and angle of the pipes were changed. All the parameters are demonstrated in Fig. 3, where θ_l, L_p, S_T and D_p are installation angle, pipe length, installation transverse distance and pipe diameter, in turn.

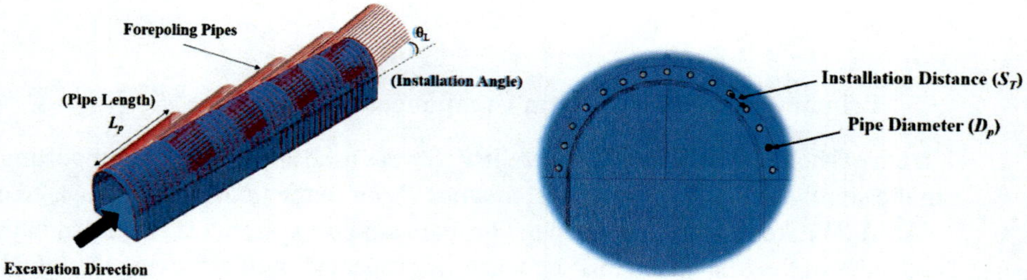

Fig. 3. Design variables of forepoling pipes.

Table 5 provided a summary of several cases that evaluated through the parametric study to capture the effect of forepoling pipe parameters on controlling the induced settlements. To evaluate the influence of changing each parameter's value, all the other parameters were considered constant (the values related to the case study) while that specific value was changed. Ranges of 8 to 15 m, 6 to 15 cm, 20 to 50 cm and 3 to 12° were assigned to the pipe length, diameter, transverse distance and installation angle, respectively.

Table 5. Design schemes devised considering different variables for forepoling pipes.

Design schemes	Length (*Lp*) (m)	Diameter (*Dp*) (cm)	Installation distance (*ST*) (cm)	Installation angle (θL) (degree)
Case 1	8			
Case 2	10	9	40	9
Case 3	12			
Case 4	15			
Case 5		6		
Case 6	1	9	4	9
Case 7	2	12	0	
Case 8		15		
Case 9			20	
Case 10	12	9	30	9
Case 11			40	
Case 12			50	
Case 13				3
Case 14	1	9	4	6
Case 15	2		0	9
Case 16				12

4 Results and Discussion

According to the results of modeling the tunnel fortified with UAM and without it, UAM was found to be significantly effective in reducing the settlements of the tunnel crown and ground surface. The remarkable pressure induced by ground loss due to excavation is actually resisted by the tunnel inverts, which can justify the positive effect of UAM. The contours of the vertical displacement of the two aforementioned cases are shown in Fig. 4. It is noted that the jagged lines observed in Fig. 4(b) pertain to the deployed explicit analysis method to optimize the iteration process adopted for convergence.

Figure 5 provides a comparison of the settlement results for all cases in presence and absence of UAM. As it can be inferred from Fig. 5, the amount of settlement in both ground surface and tunnel crown is falling as the excavation proceeds to the end of 42 m of the tunnel length. In other words, the maximum settlement is experienced at the entrance portal of the tunnel which makes it the critical point for evaluating the effect of different parameters. Based on the plot, UAM could decrease the vertical settlement of the tunnel crown and ground surface from 166 mm to 40 mm and 62 mm to 36 mm, equivalent to 76% and 42% of settlement reduction, correspondingly.

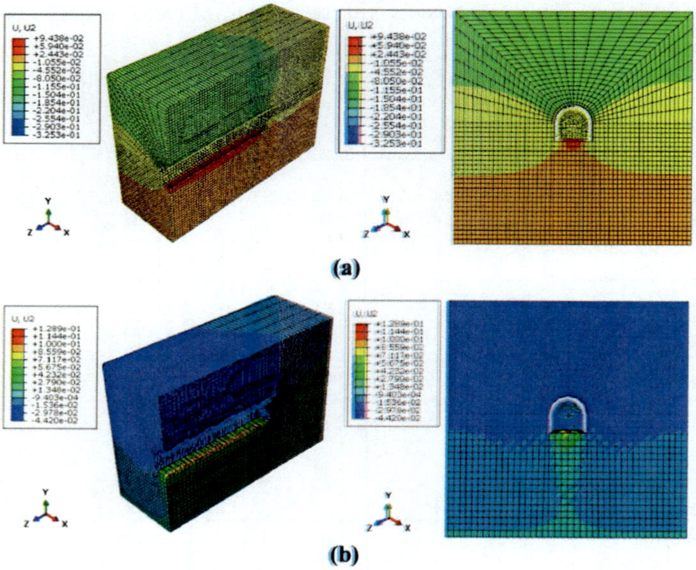

Fig. 4. Vertical displacement contours, in the case (a) without UAM application, and (b) with UAM application.

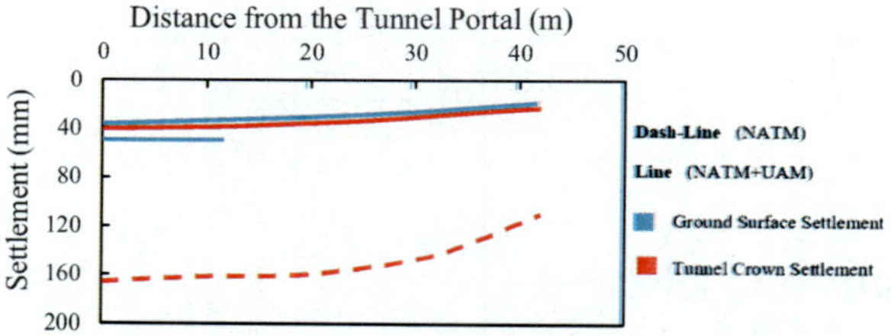

Fig. 5. Results of tunnel excavation-induced settlements for the cases in presence and absence of UAM.

For assessing the influence of changing the value of each parameter of the forepoling pipes, a series of parametric analyses were carried out. In each analysis, the vertical displacements of the tunnel crown have been plotted for various values of the pertinent parameter, as demonstrated in Fig. 6. As the changes in the settlements at tunnel portal were negligible after 42 m of excavation, the plots show the curves of displacement for the same amount (the first 42 steel frames). As evidenced in the plots, settlement values decrease noticeably by increasing the pipe diameter from 6 to 12 cm (Fig. 6a), increasing the pipe length from 8 to 12 m (Fig. 6b), decreasing the installation angle from 12 to 6° (Fig. 6c) and decreasing the installation distance from 60 to 30 cm (Fig. 6d). The reduction of settlement resulted by increasing the pipe diameter is because of the growth induced in the section moment of inertia which can lead to a surge in bending capacity of the forepoling pipes. Similarly, the fact that can justify the reduction of displacements by reducing the installation angle is that the distance from the umbrella to the tunnel crown is lessened and accordingly the capability of the UAM in transferring the applied stresses is improved.

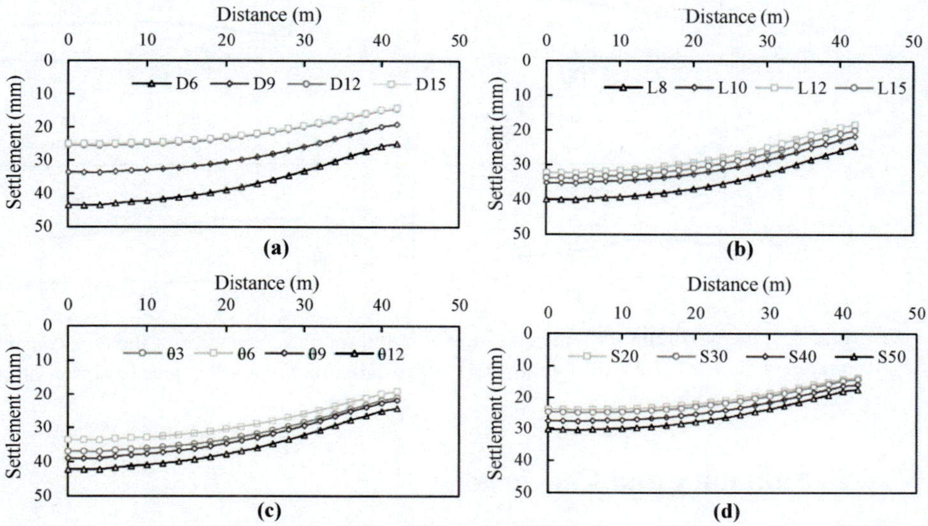

Fig. 6. Tunnel crown vertical displacements along the tunnel axis for various values of the pipes design parameters: (a) Diameter (Dp), (b) Length (Lp), (c) Installation angle (θL), and (d) Installation distance (ST).

It is of utmost importance to find the optimum values of each parameters to control the settlements induced in the tunnel crown effectively while avoiding over estimating the dimensions and installation-related factors. To achieve this end, the authors plotted the maximum settlements occurred at the tunnel entrance versus the amount of each parameter for the analyzed cases, all shown in Fig. 7. In all the plots of the Fig. 7, there is a point after which any further change in the parameter contributes to a negligible reduction of the tunnel crown settlement. Hence these points can be considered as the optimum values of the design factors. The optimum values of diameter, length, installation angle and distance of the forepoling pipes based on Fig. 7(a) through (d) are 12 cm, 12 m, 6° and 30 cm, respectively. It is an important point that for the studied cases with tunnel diameter (D_{tunnel}) of 8 m, the optimum length of the pipes is 12, which is 1.5 D_{tunnel}.

It worth noting that regarding the ground surface settlements, similar analysis results were obtained associated with the influence of the pipes design parameters on controlling the ground surface settlements. Essentially, the overall patterns of the variations of ground surface settlements indicated similar trends for optimized design parameters of the pipes to those suggested by the analysis carried out for the tunnel crown vertical displacements (Morovatdar et al. 2020a).

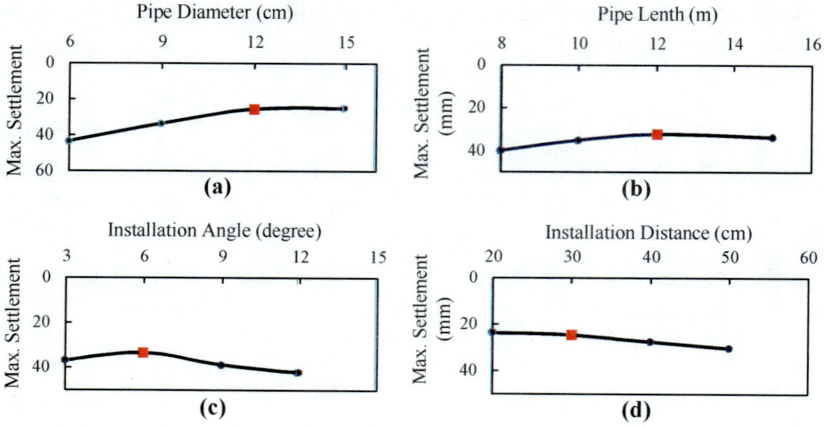

Fig. 7. The maximum settlements at tunnel portal for various values of the pipes design parameters: (a) Diameter (*Dp*), (b) Length (*Lp*), (c) Installation angle (θL), and (d) Installation distance (*ST*).

5 Summary and Conclusions

Safe construction of the shallow-buried tunnels in soft grounds, in which the tunnels might be located at the vicinity of the densely populated regions, is of utmost importance. Hence, deploying a high performance method to stabilize the tunnel and prevent the potential excavation induced settlement is a critical issue for researchers and practitioners. In this paper, a comprehensive numerical study is conducted on the UAM which is one of the tunnel supporting systems. The 3D numerical finite element modeling of the cases including a case study approved that when employing UAM especially in soft soils, the vertical settlements occurred in the crown of tunnel and on the ground surface dropped by 76% and 42%, respectively. This remarkable reduction confirms the beneficial impacts of UAM in tunnel settlement mitigation.

Subsequently, various design parameters in terms of the dimensions of the forpoling pipes and installation layout factors, were extensively evaluated by numerical modeling of several cases in ABAQUS/CAE. Generally, a descending trend was observed in the induced settlement along the tunnel axis when increasing the dimensions of the pipes and reducing the angle and transverse distance of the installation of the pipes. However in each variable evaluation, the settlements reduced up to a specific point and after that point, altering the corresponding variable did not improve the stability of the tunnel in terms of decreasing the crown settlements. Thus, that point and the variable value corresponding to it was considered as the optimum value in design procedure of the pipes and installation layout decisions. For the range of the design inputs evaluated in this study, the optimum values for pipes length, diameter, installation angle and installation distance were obtained as 12 m, 12 cm, 6 degrees and 30 cm. comparing the optimized values of the pipes design variables from numerical simulations with those extracted from the design plans of the Ghazvin-Rasht project has yielded identical results.

For the future developments and enhancements it is suggested to evaluate the influence of earthquake on the efficiency of the UAM-reinforced tunnels under dynamic loading conditions. Liquefaction is another major phenomenon that needs to be incorporated into the analysis of the stability of the tunnels constructed in soft grounds with weak geotechnical properties.

References

Aksoy, C.O., Onargan, T.: The role of umbrella arch and face bolt as deformation preventing support system in preventing building damages. Tunn. Undergr. Space Technol. **25**(5), 553–559 (2010)

Alimohammadi, H., Abu-Farsakh, M.: Finite element parametric study on rutting performance of geosynthetic reinforced flexible pavements. Transp. Res. Board **98** (2019)

Beizaei, M., Hosseininia, S.E., Morovatdar, A.: Investigating the practical conditions to utilize brick stair wall method as a supporting structure in urban excavation. In: Geo-Congress 2020: Engineering, Monitoring, and Management of Geotechnical Infrastructure, pp. 429–439. Reston, VA: American Society of Civil Engineers (2020)

Beyzaei, M., Hosseininia, S.E.: A numerical investigation on the performance of the brick stair wall as a supporting structure by considering adjacent building. KSCE J. Civil Eng. **23**(4), 1513–1521 (2019)

Dehghani, N.L., Rahimi, M., Shafieezadeh, A., Padgett, J.E.: Parameter estimation of a fractional order soil constitutive model using KiK-Net downhole array data: a Bayesian updating approach. In: Geo-Congress 2020 Earthquake Engineering and Soil Dynamics (2020)

Simulia, D.S.: Abaqus/CAE User's Manual. SIMULIA, Providence (2007)

Estahbanati, S.H., Boushehri, R., Soroush, A., Ghasemi-Fare, O.: Numerical analysis of dynamic response of lifelines facilities adjacent to deep excavations. In: Geo-Congress 2020: Geotechnical Earthquake Engineering and Special Topics, pp. 253–265. Reston, VA: American Society of Civil Engineers (2020)

Farhangi, V., Karakouzian, M.: Design of bridge foundations using reinforced micropiles. In: International Road Federation Global R2T Conference & Expo, pp. 78–83 (2019)

Farhangi, V., Karakouzian, M., Geertsema, M.: Effect of Micropiles on clean sand liquefaction risk based on CPT and SPT. Appl. Sci. **10**, 3111 (2020). https://doi.org/10.3390/app10093111

Klotoé, C.H., Bourgeois, E.: Three dimensional finite element analysis of the influence of the umbrella arch on the settlements induced by shallow tunneling. Comput. Geotech. **110**, 114–121 (2019)

Morovatdar, A., Palassi, M., Ashtiani, R.S.: Effect of Pipes characteristics in umbrella arch method on controlling the tunneling-induced settlements in soft grounds. J. Rock Mech. Geotech. Eng. (2020). https://doi.org/10.1016/j.jrmge.2020.05.001

Morovatdar, A., Palassi, M., Beizaei, M.: Evaluation of the efficiency of the umbrella arch method in urban tunneling subjected to adjacent surcharge loads. In: Geo-Congress 2020: Engineering, Monitoring, and Management of Geotechnical Infrastructure, pp. 258–268, Reston, VA: American Society of Civil Engineers (2020b)

Morovatdar, A., Ashtiani, R.S., Licon, C., Tirado, C., Mahmoud, E.: Novel framework for the quantification of pavement damages in the overload Corridors. Transportation Research Record (TRR). J. Transport. Res. Board (2020c). https://doi.org/10.1177/0361198120925807

Morovatdar, A., Ashtiani, R.S., Licon, C., Tirado, C.: Development of a mechanistic approach to quantify pavement damage using axle load spectra from south Texas overload corridors. In: Geo-Structural Aspects of Pavements, Railways, and Airfields Conference (GAP 2019), Colorado Springs, CO, USA (2019)

Ocak, I., Selcuk, E.: Comparison of NATM and umbrella arch method in terms of cost, completion time, and deformation. Arab. J. Geosci. **10**(7), 1–11 (2017). https://doi.org/10.1007/s12517-017-2938-8

Ocak, I.: Control of surface settlements with umbrella arch method in second stage excavations of Istanbul Metro. Tunn. Undergr. Space Technol. **23**(6), 674–681 (2008)

Oke, J., Vlachopoulos, N., Diederichs, M.S.: Numerical analyses in the design of umbrella arch systems. J. Rock Mechan. Geotech. Eng. **6**(6), 546–564 (2014)

Oke, J., Vlachopoulos, N., Diederichs, M.S.: Semi-analytical model for umbrella arch systems employed in squeezing ground conditions. Tunn. Undergr. Space Technol. **56**, 136–156 (2016)

Qian, Z., Zou, J., Pan, Q., Dias, D.: Safety factor calculations of a tunnel face reinforced with umbrella pipes: a comparison analysis. Eng. Struct. **199**, 109639 (2019)

Ranjbarnia, M., Rahimpour, N., Oreste, P.: A simple analytical approach to simulate the arch umbrella supporting system in deep tunnels based on convergence confinement method. Tunn. Undergr. Space Technol. **82**, 39–49 (2018)

Salimi, K., Ghazavi, M.: Soil reinforcement and slope stabilisation using recycled waste plastic sheets. Geomechan. Geoeng. **1**, 1–2 (2019)

Salimi, K., Cerato, A.B., Vahedifard, F., Miller, G.A.: A general model for tensile strength characteristic curve of unsaturated soils. J. Geotech. Geoenviron. Eng. **147**, 9348 (2020)

Salimi, K., Cerato, A.B., Vahedifard, F. and Miller, G.A.: Tensile strength of compacted clays during desiccation under elevated temperatures. ASTM Geotech. Testing J. (2020a)

Satvati, S., Alimohammadi, H., Rowshanzamir, M., Hejazi, S.M.: Bearing capacity of shallow footings reinforced with braid and geogrid adjacent to soil slope. Int. J. Geosynth. Ground Eng. **6**(4), 1–12 (2020). https://doi.org/10.1007/s40891-020-00226-x

Schumacher, F.P., Kim, E.: Modeling the pipe umbrella roof support system in a Western US underground coal mine. Int. J. Rock Mech. Min. Sci. **60**, 114–124 (2016)

Song, K.I., Cho, G.C., Chang, S.B., Lee, I.M.: Beam–spring structural analysis for the design of a tunnel pre-reinforcement support system. Int. J. Rock Mech. Min. Sci. **59**, 139–150 (2013)

Wittke, W., Pierau, B., Erichsen, C.: New Austrian tunneling method (NATM)-stability analysis and design. WBI Essen. **5**, 2441 (2006)

Zamanian, S., Hur, J., Shafieezadeh, A.: Significant variables for leakage and collapse of buried concrete sewer pipes: a global sensitivity analysis via Bayesian additive regression trees and Sobol'indices. Struct. Infrast. Eng. **14**, 1–13 (2020)

Influence Factors of the Adhesion Strength of Clayey Soil

Shuying Wang, Pengfei Liu(✉), and Jiazheng Zhong

School of Civil Engineering, Central South University, Changsha 410075, Hunan, China
pengfeiliu@csu.edu.cn

Abstract. When earth pressure balance (EPB) shield tunnels are constructed through clayey ground, the soil adheres to the cutter and cutterhead due to the high adhesion strength between the steel and the clay. To investigate the influence of different factors on the adhesion strength, this study used montmorillonite, kaolin and mixtures of the two as test soils. The adhesion strength between steel and clay was determined with a customized rotary shear apparatus. The results show that the adhesion strength between the steel and the clay increased with the consistency index. As the consistency index decreased, the effect of the normal pressure on the adhesion strength gradually weakened. As the contact angle of the shear plate increased, the adhesion strength decreased. The plasticity index had little effect on the adhesion strengths. The adhesion strength increased gradually with increasing surface roughness. The most important factors which influenced the adhesion strength are the soil consistency index, the normal pressure and the contact angle of the shear plate.

Keywords: EPB shield · Clayed soil · Adhesion strength

1 Introduction

The earth pressure balance (EPB) shield has been widely used in urban metro construction due to its high efficiency and safety (Herrenknecht et al. 2011; Zumsteg et al. 2012; Ye et al. 2016; Liu et al. 2018). When the EPB shield crosses a soil layer with a high clay mineral content, the muck may adhere to the cutterhead and cutters. The EPB shield becomes clogged and is not able to tunnel.

The main reason for shield clogging is the high adhesion strength between steel and clay. However, the adhesion between steel and clay is a complex problem and is influenced by several factors, such as the clay state, clay water content, normal pressure, and steel surface state. Some researchers have conducted related studies. Sass and Burbaum (2009) studied the effects of material type, steel surface roughness and normal pressure on the adhesion strength of clay on different materials using a customized apparatus. Spagnoli (2011) and Feinendegen et al. (2010) determined the variation in the adhesion of clay with the consistency index I_c by cone pull-out testing. Liu et al. (2018) reported that the adhesion strength first increased and then decreased with an increase in the consistency index of clay, and the adhesion strength reached a maximum value when

Y. Liu et al. (Eds.): GeoChina 2021, SUCI, pp. 57–64, 2021.
https://doi.org/10.1007/978-3-030-80316-2_6

the consistency index was approximately 1. Azadegan and Massah (2012) reported that when the temperature increased from 5 °C to 30 °C, the adhesion strength decreased by 82%. The surface tension of the moisture in the soil decreased continuously with an increase in temperature, resulting in a decrease in adhesion strength. Zumsteg and Puzrin (2012) used a customized shear plate apparatus to investigate the effect of roughness, water content and normal pressure on the adhesion strength between steel and clay and concluded that the adhesion strength increased with the roughness of the steel surface and the normal pressure and that the relationship between the adhesion strength and the water content was logarithmic.

Thus, the adhesion strength between steel and clay can be affected by the clay type, the soil consistency index, the metal hydrophilicity, the surface roughness and the normal pressure. However, the existing research does not fully explore the impact of various factors on adhesion strength. The paper investigated the variation in adhesion strength between steel and clay was investigated considering the consistency index, normal pressure, plasticity index, contact angle and plate surface roughness using a rotary shear apparatus.

2 Testing Materials and Approaches

2.1 Testing Materials

The testing soils contained montmorillonite, kaolin and mixtures of the two. Montmorillonite and kaolin were mixed at mass ratios of 3:1 (M_1), 1:1 (M_2) and 1:3 (M_3). The Atterberg limits of the tested soils are shown in Table 1.

Table 1. The Atterberg limits of the clays

Tested soil	Plastic limit (%)	Liquid limit (%)	Plasticity index
Montmorillonite	59.9	316.8	256.9
Mixture 1 (M_1)	52.5	254.0	201.5
Mixture 2 (M_2)	34.3	156.7	122.4
Mixture 3 (M_3)	32.4	98.0	65.6
Kaolin	31.8	56.2	24.4

The tested metal plate was made of 303 stainless steel with a surface roughness of 0.34 μm. When the influence of the surface roughness was investigated, the adhesion strength was determined using shear plates made of 303 stainless steel with surface roughnesses of 0.34, 4.84 and 11.03 μm.

The contact angle θ refers to the angle between the tangent line of the gas-liquid interface and the solid-liquid boundary line at the intersection of a gas, liquid and solid. When the liquid is water, the contact angle characterizes the hydrophilicity of the solid material. The contact angle is related to the material properties. When the influence of the contact angle was investigated, the adhesion strength was determined with shear plates made of Cr12MoV mold steel, 45 steel, Q235 steel, 303 stainless steel, polyoxymethylene (plastic) and polyethylene (plastic). The contact angles of the steels are shown in Table 2.

Table 2. Parameters of the shear plates

Material	Contact angle (°)
Cr12MoV mold steel	58.96
45 steel	58.76
Q235 steel	57.15
304 stainless steel	62.04
Polyoxymethylene (POM)	78.32
Polyethylene (PE)	94.51

2.2 Rotary Shear Test

The adhesion strength between the steel and clay under different normal pressures was measured using a customized rotary shear apparatus, as shown in Fig. 1 (Liu et al. 2018). The rotary shear apparatus contains a motor, specimen box, loading device and shear plate. Soils with a consistency index of less than 0.4 are too soft to ensure the pressure tightness of the screw conveyor, whereas soils with a consistency index > 0.75 are too stiff to be used as a supporting medium (Thewes and Budach 2010; Hollmann and Thewes 2013). Thus, the range of the consistency index values of the soil specimens was appropriately expanded and was determined to be 0.1 to 1.0, at which the adhesion strength reached the maximum value (Liu et al. 2018). The testing conditions are listed in Table 3.

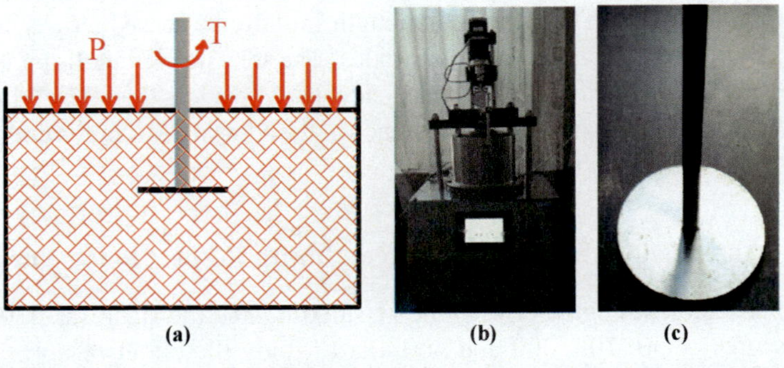

Fig. 1. Rotary shear apparatus: (a) schematic diagram, (b) photo of the apparatus and (c) shear plate

Table 3. Overview of the testing conditions

Parameter	Values
Consistency index	0.1, 0.2, 0.3, 0.4, 0.5, 0.6, 0.8, 1.0
Normal pressure (kPa)	100, 200, 300, 400
Plasticity index (I_p)	24.4, 122.4, 256.9
Contact angle (°)	57.15, 58.67, 58.96, 62.04, 78.32, 94.51
Surface roughness (μm)	0.34, 4.84, 11.03

The test procedures were as follows. The box of the rotary shear apparatus was filled with soil in layers. Then, the top plate was positioned on the soil specimen. A certain normal pressure was applied to expel the air between the top plate and the soil specimen. Then the exhaust valves of the test equipment were closed and the desired pressure was applied to the specimen. After the normal displacement of the top plate of the box became stable, the shear plate was rotated at a velocity of 20°/min. The torque was recorded, and the adhesion strength was calculated with formula (1). The detailed test procedures of the rotary shear test can be found in Liu et al. (2019).

$$a = \frac{6T}{\pi D^3} \tag{1}$$

where a - the adhesion stress between the clay and steel, T - the torque measured by the torque sensor, and D - the diameter of the steel plate.

3 Test Results

3.1 Variation in the Adhesion Strength with Respect to the Consistency Index

The adhesion strength between the M_2 mixture and the 303 stainless steel was determined at different consistency index values (Fig. 2). The normal pressure was set to 50, 100, 200 and 300 kPa. The adhesion strength initially increased slowly and then increased rapidly with increasing consistency index. The adhesion strengths under different normal pressures were similar as the consistency index was less than 0.6. While the consistency index was higher than 0.6, the difference in the adhesion strengths under different normal pressures increased markedly.

3.2 Variation in the Adhesion Strength with Respect to the Normal Pressure

The adhesion strength between the M_2 mixture and the steel was determined at normal pressures of 50, 100, 200 and 300 kPa to study the effect of normal pressure on the adhesion strength. The consistency index (I_c) values of the soil specimens were set to 0.4 and 0.8, as shown in Fig. 3. When the consistency index was 0.8, the increase in the adhesion strength decreased with an increase in the normal pressure. When the consistency index was 0.4, the adhesion strength remained almost constant with increasing

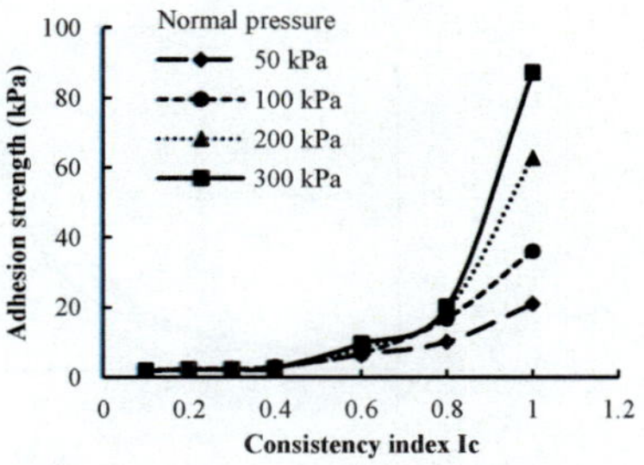

Fig. 2. Variation in the adhesion strength with the consistency index under the normal pressures of 50, 100, 200 and 300 kPa

normal pressure. Since the soil specimen with a consistency index of 0.4 was almost saturated, the normal pressure was supported by the pore water in the soil sample. Thus the normal pressure has little effect on the adhesion strength. As the consistency index increased, the soil saturation decreased. The normal pressure would have a significant influence on the adhesion strength.

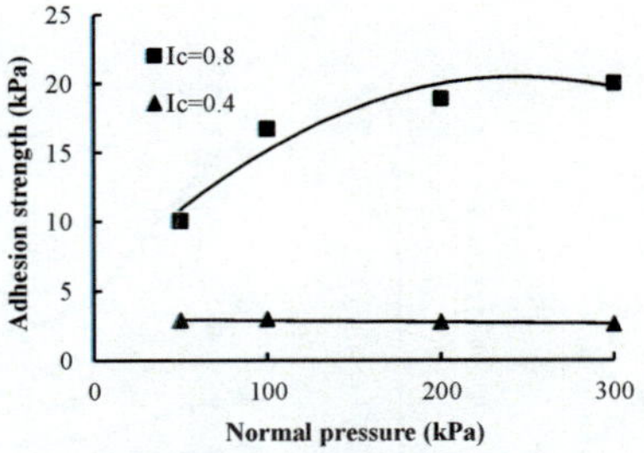

Fig. 3. Variation in the adhesion strength with respect to normal pressure

3.3 Variation in the Adhesion Strength with Respect to the Plasticity Index

The adhesion strengths between the steel and the montmorillonite, kaolin, and the M_2 mixture were determined at the normal pressure of 100 kPa to study the effect of the plasticity index I_p on the adhesion strength (Fig. 4). No major difference was observed in the adhesion strengths among the clays at the same consistency index value. This result indicated that the plasticity index had a limited effect on the adhesion strength if the soil consistency states were similar.

3.4 Variation in the Adhesion Strength with Respect to the Contact Angle

To study the effect of contact angle on adhesion strength, shear plates made of the Cr12MoV mold steel, 45 steel, Q235 steel, 303 stainless steel, polyoxymethylene and polyethylene were tested. The adhesion strength between the different shear plates and the clay was measured at a normal pressure of 100 kPa (Fig. 5). The adhesion strengths of clay decreased with an increase of contact angles. Since the contact angles of the

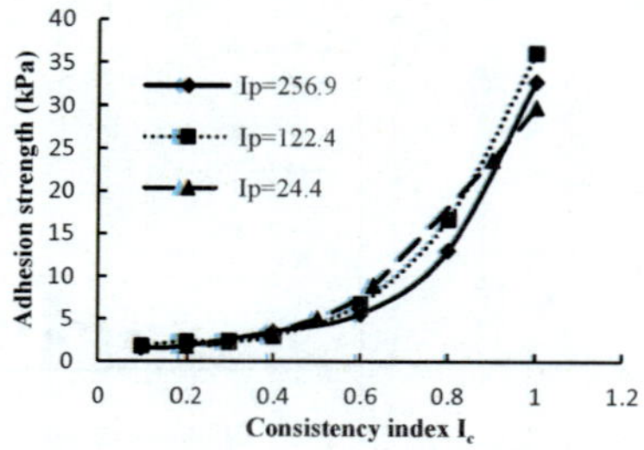

Fig. 4. Variation in the adhesion strength with respect to the consistency index for different clays under a normal pressure of 100 kPa

tested metal materials ranged from 57.15° to 62.04°, there were no large differences in the adhesion strength among the different metal materials. However, the adhesion strengths between the two organic plastics and the clay were significantly lower than those between the metals and the clay due to the much higher contact angles of the plastics. Therefore, the effect of contact angle on adhesion strength is great; however, the metal materials do not have different adhesion strengths due to their similar contact angles.

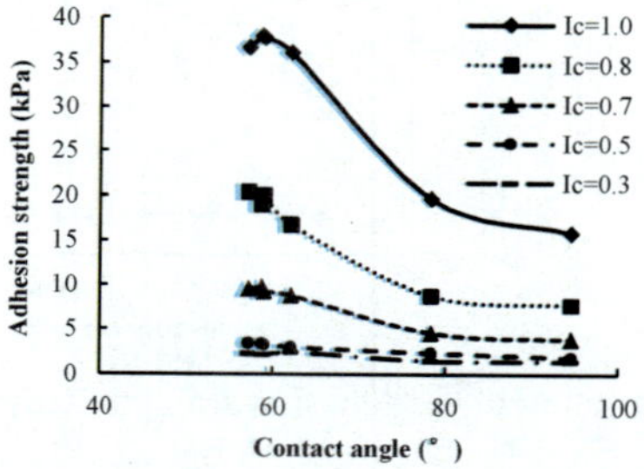

Fig. 5. Variation in the adhesion strength with respect to contact angle under a normal pressure of 100 kPa

3.5 Variation in the Adhesion Strength with Respect to the Surface Roughness

To study the effect of surface roughness on the adhesion strength, the adhesion strengths between the M_2 clay mixture and the steel plates with surface roughnesses of 0.34, 4.84 and 11.03 μm were determined at normal pressures of 50 and 100 kPa (Fig. 6). The adhesion strengths of the shear plates with different surface roughnesses remained basically constant as the consistency index was low. At higher consistency index values, the adhesion strength increased with the surface roughness. However, the effect of surface roughness on the adhesion strength was generally not significant.

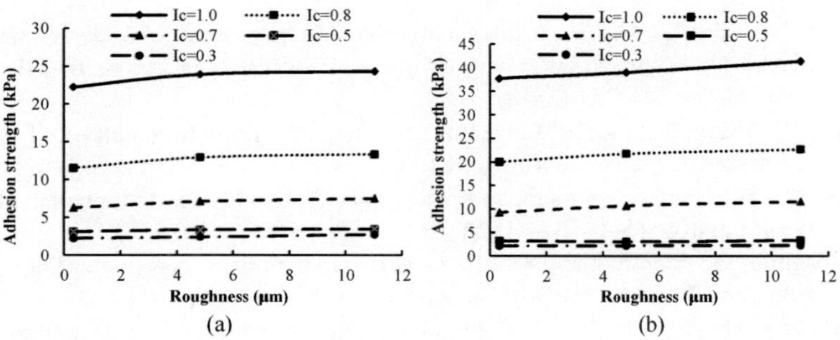

Fig. 6. Variation in adhesion strength with respect to surface roughness at different consistency index values under normal pressures of (a) 50 kPa and (b) 100 kPa

4 Conclusions and Prospects

When the consistency index of the soil specimen was less than 1, the adhesion strength between the steel and the clay increased with an increase in the consistency index. As the consistency index decreased (i.e., as the water content and saturation of the soil specimens increased), the effect of the normal pressure on the adhesion strength gradually weakened. As the contact angle of the shear plate increased, the hydrophilicity gradually decreased, and the adhesion strength decreased. When the soil specimens with different plasticity index values had the same consistency index, the resulting adhesion strengths were similar. The adhesion strength increased gradually with an increase in surface roughness. The most important factors which influenced the adhesion strength are the soil consistency index, the normal pressure and the contact angle of the shear plate.

As stated in the paper, the hydrophilicity of the cutterhead and cutters in the EPB has significant effect on the adhesion strength. Thus, how to make hydrophobic cutting tools is a key technology to avoid EPB clogging.

Acknowledgments. The financial support from the National Natural Science Foundation of China (No. 51778637) and the National Key R&D Program of China (No. 2017YFB1201204) are acknowledged and appreciated.

References

Azadegan, B., Massah, J.: Effect of temperature on adhesion of clay soil to steel. Cercet. Agron. Moldova **45**(2), 21–27 (2012)

Feinendegen, M., Ziegler, M., Spagnoli, G.: A new laboratory test to evaluate the problem of clogging in mechanical tunnel driving with EPB-shields. In: ISRM International Symposium-Eurock (2010)

Herrenknecht, M., Thewes, M., Budach, C.: The development of earth pressure shields: from the beginning to the present/Entwicklung der Erddruckschilde: Von den Anfängen bis zur Gegenwart. Geomech. Tunnel. **4**(1), 11–35 (2011)

Hollmann, F.S., Thewes, M.: Assessment method for clay clogging and disintegration of fines in mechanised tunnelling. Tunnel. Undergr. Space Technol. Incorp. Trenchless Technol. Res. **37**(13), 96–106 (2013)

Liu, P., Wang, S., Ge, L., et al.: Changes of Atterberg limits and electrochemical behaviors of clays with dispersants as conditioning agents for EPB shield tunnelling. Tunn. Undergr. Space Technol. **73**, 244–251 (2018)

Liu, P., Wang, S., Shi, Y.: Tangential adhesion strength between clay and steel for various soil softnesses. J. Mater. Civil Eng. **31** (2019)

Sass, I., Burbaum, U.: A method for assessing adhesion of clays to tunneling machines. Bull. Eng. Geol. Environ. **68**(1), 27–34 (2009)

Spagnoli, G.: Electro-chemo-mechanical manipulations of clays regarding the clogging during EPB-tunnel driving. RWTH Aachen University (2011)

Thewes, M., Budach, C.: Soil conditioning with foam during EPB tunnelling. Geomech. Tunnel. **3**(3), 256–267 (2010)

Ye, X., Wang, S., Yang, J.: Soil conditioning for EPB shield tunneling in argillaceous siltstone with high content of clay minerals: case study. Int. J. Geomech. (2016). https://doi.org/10.1061/(ASCE)GM.1943-5622.0000791

Zumsteg, R., Plötze, M., Puzrin, A.M.: Effect of soil conditioners on the pressure and rate-dependent shear strength of different clays. J. Geotech. Geoenviron. Eng. **138**(9), 1138–1146 (2012)

Zumsteg, R., Puzrin, A.M.: Stickiness and adhesion of conditioned clay pastes. Tunnel. Undergr. Space Technol. Incorp. Trenchl. Technol. Res. **31**(5), 86–96 (2012)

Numerical Investigation on Cracking Behavior of Shield Tunnel Lining Subjected to Surface Loading: A Parametric Study

Jiachong Xie, Jinchang Wang(✉), Weiming Huang, Zhongxuan Yang, and Rongqiao Xu

Department of Civil Engineering, Zhejiang University, Hangzhou 310058, China
wjc501@zju.edu.cn

Abstract. An optimized finite element method (FEM) based on layer structure method has been put forward, which combined the nonlinear behavior of both the structure and the soil surrounding. The lining is simulated with solid element and meshed finely, while the longitudinal joint between segments is modeled by 6-node line interface element. Nonlinear mechanical behaviors of soil, concrete, reinforcement and interfaces are considered with advanced material models. A parameter study is conducted to reveal the influence of several concerning parameters such as the rotation angle of the ring, the width of surface loading, the offset of surface loading and the depth of the tunnel. To better illustrate the result, several indexes including maximum crack width, crack index and crack ratio is introduced to evaluate and forecast crack behavior. The results show that the crack-resistant ability of the tunnel is slightly enhanced as the rotation angle of the ring increases. A more unfavorable influence can be brought to linings by the shift of the surface loading when its offsetting is smaller than 30 m in the case. However, the influence on segment cracking caused by the change of the tunnel depth is complicated when stress redistribution after tunnel excavation is considered, and there is an unfavorable buried depth when the crack ratio reaches maximum. Finally, a simple fitting equation, which is based on the previously mentioned numerical simulation, is provided to establish the relationship between the horizontal ovalisation and the crack ratio.

1 Introduction

Linked by bolts and rubber packers, shield tunnel segments are assembled into a massive structure with a certain degree of flexibility, while it has quite a strong resistance of oval-shaping and cracking. Even if owning excellent working performance, the shield tunnel and its segments still suffer from overrunning ovalisation and other diseases because of intense environmental disturbances such as adjacent excavation, surface loading and tunnel under-crossing during operation. Among them, unexpected surface loading in the proximity of tunnel may be the most common one. It is generally accepted that the tunnel structure in soft soil areas is more sensitive to the surface loading. Besides of the ovalisation and settlement of the tunnel, diseases like concrete cracking, water leakage and segment misalignment appear frequently within the affected subway section

Y. Liu et al. (Eds.): GeoChina 2021, SUCI, pp. 65–79, 2021.
https://doi.org/10.1007/978-3-030-80316-2_7

(Shao et al. 2016). Thus, it is of great significance for practical engineering to analyze the deformation features and cracking characteristics of concrete segments under the condition of surface loading.

Extensive studies have shown that the mechanical behavior of lining segments presents obvious nonlinearity (Arnau and Molins 2012). Among these studies, the nonlinear response of lining was usually analyzed by the load structure method where the soil springs were used to simulate the interaction between the tunnel structure and soil when the corresponding structural response was solved under various pre-assumed loading condition. Zhou et al. (2019) proposed a multi-scale simulation method which considered three levels of material, segment and tunnel, and the nonlinear response of the tunnel segment under different ovalisation conditions was analyzed. Based on ADINA software, Chen and Mo (2009) established a refined shield tunnel segment model to elaborately consider bolts and hand holes, and the nonlinear response of the structure under shield thrust and torsion loading were studied. Wang and Zhang (2013) simulated the segments and bolts of the tunnel with solid elements, and the deformation mechanism of the shield tunnel under overloading was analyzed. Based on the numerical analysis of load structure method, a quick insight into the structural damage response can be obtained. However, the simplified soil spring and pre-assumed loading cannot reflect the real lining-ground interaction. Consequently, the authenticity of the simulation results is in doubt.

Another calculation strategy—layer structure method is more preferable to be adopted to involve complex geotechnical conditions such as more realistic lining interaction. In this method, the lining structure, the surrounding stratum and its contact surface are all modeled in detail at the same time. However, to yield a solution, it is necessary to limit the scale of model and the precision meshwork of the FEM model. Papanikolaou and Kappos (2014) studied the structural response and cracking behavior of a horseshoe-shaped tunnel based on ATENA software under various conditions such as fire, blast and earthquake with a plane strain model. Xu et al. (2019) put forward a new numerical approach for cracked segment to analyze the mechanical behavior of lining with multi-cracks, and the cracking mechanism of segmental linings investigated from on-site inspection was revealed by the numerical results.

In addition, some researchers (Sun et al. 2017; Yang et al. 2016) combined the above two mentioned calculation strategies in their studies. Firstly, layer structure method with a homogeneous ring having degraded bending stiffness was adopted to calculate the soil pressure acting on the lining or tunnel ovalisation deformation. Then, the above calculated results were treated as input parameters in the subsequent step which was carried out in the framework of load structure method. In this way, soil pressure acting on the lining and nonlinear response of lining structure are calculated successively and separately. But there are still shortages such as computational complexity and nonobjective simulation on the coupling effect of surrounding soil and linings.

The motivation of the present study is to propose an optimized finite element modelling method in the framework of layer structure method to involve the nonlinear behavior of both the surrounding soil and segments with advanced material models. A two-dimensional model of a real tunnel section of Hangzhou Metro Line 1 is established, where hardening soil model with small strain (HS-small model) and total strain crack model for reinforced concrete are adopted. A parameter study is performed in order to reveal the influence of the rotation angle of the tunnel, surface loading width, the offset of surface loading and the depth of the tunnel. Based on it, the crack characteristics of tunnel segments under different loading condition are discussed, which can provide an instructive scientific support for the protection and maintenance of existing shield tunnels.

2 Modeling Strategies

To achieve a comprehensively modelling, commercial software DIANA 10.3 was used in the present study. It takes in plenty of advanced material models, while providing a feasible and sufficient numerical tool and making it possible to model the lining and surrounding soil as well as their interaction in detail.

The case is from Hangzhou Metro Line 1 (Xie et al. 2020). The tunnel lining is made with C50 reinforce concrete, and the internal and external diameters of the ring are equal to 5.5 m and 6.2 m respectively. Thus, the lining thickness is equal to 350 mm. The whole ring is divided into 1 key segment (F), 2 adjacent segments (L1, L2) and 3 standard segments (B1, B2, B3). And every ring has a width of 1.2 m, with a rotation angle of 45°between adjacent rings, as shown in Fig. 1. Taking the requirement of convergence for nonlinear analysis into account, the modeling is carried out in two-dimensional and the strengthening effect of the circumferential joint is ignored, which is a more unfavorable case.

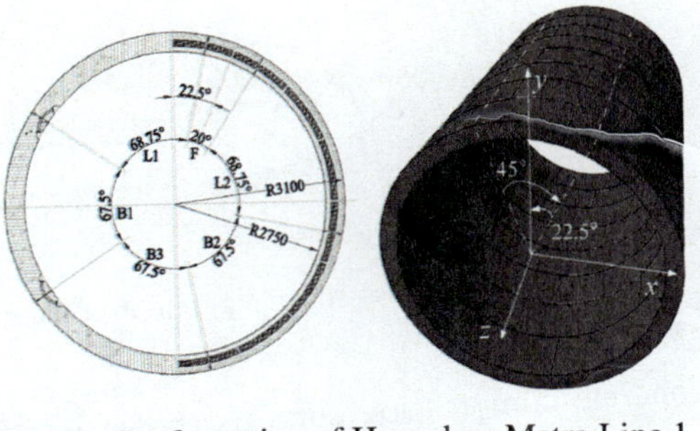

Fig. 1. Configuration of Hangzhou Metro Line 1

2.1 Concrete Behavior

Refer to the local crack direction, the total strain crack model can be divided into rotating, fixed and mixed crack modes. The rotating crack model is adopted in present paper for it was proved to have a better simulation result for reinforced concrete structures (Hendriks et al. 2017). The parameters of segment concrete are shown in Table 1. The exponential model is adopted to describe the tensile softening behavior of concrete, the stress-strain curve is shown in Fig. 2(a), in which h_{cr} is band width which can be calculated fro element size. G_f^{I} is suggested to derive from the equation (Hendriks et al. 2017):

$$G_f^I = 73 \times f_{cm}^{0.18}$$

Where f_{cm} is mean compression strength, herein f_{cm} equals to 58 MPa.

Table 1. Parameters of C50 concrete

Property	Value
Young's modulus, E(MPa)	34300
Poisson's ratio,v	0.2
Tensile strength, f_t(MPa)	2.64
Compression strength, f_c(MPa)	32.4
Tensile fracture energy, G_f^{I}(N/m)	151
Compression fracture energy, G_c(N/m)	37750

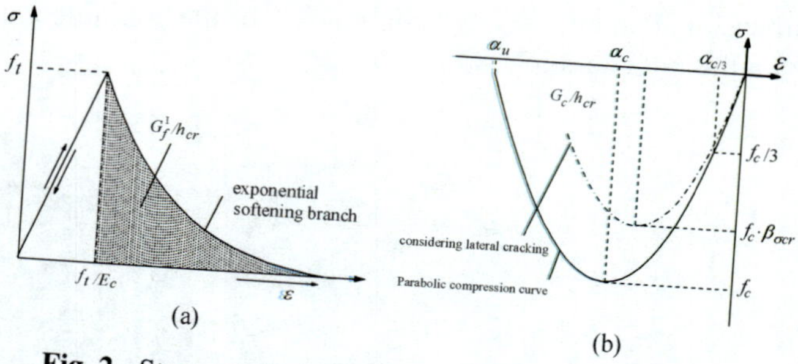

Fig. 2. Stress-strain diagram in tension and compression

The concrete compression curve is shown in Fig. 2(b), and the equation of the parabolic compression curve is defined as follows (Feenstra 1993):

$$\alpha_{c/3} = -\frac{1}{3}\frac{f_c}{E} \tag{2}$$

$$\alpha_c = -\frac{5}{3}\frac{f_c}{E} = 5\alpha_{c/3} \tag{3}$$

$$\alpha_u = \min\left(\alpha_c - \frac{3}{2}\frac{G_c}{hf_c}, 2.5\alpha_c\right) \tag{4}$$

In which α_u is ultimate compressive strain of concrete; α_c is strain at f_c; $\alpha_{c/3}$ is strain at $f_c/3$, when the concrete is in the elastic state.

Considering the fact that the tensile strain in the larger principal strain direction will cause a reduction in the compressive strength of the element perpendicular to it, which is called the compression softening effect, a reduction factor $\beta_{\sigma cr}$ is adopted to describe the behavior of lateral cracking. $\beta_{\sigma cr}$ is defined by Vecchio and Collins (1993) as follows:

$$\beta_{\sigma cr} = \frac{1}{1 + K_c} \leq 1 \tag{5}$$

$$K_c = 0.27\left(\frac{\varepsilon_1}{\alpha_c} - 0.37\right) \tag{6}$$

2.2 Reinforcement Element

The reinforcement is simplified by the grid element, which can be well embedded in the solid element. The grid element does not participate in mesh division and does not have degrees of freedom as well. It only contributes stiffness by combining with the concrete element. In 2D numerical model, it is performed as a curve. The grid element requires that the actual complex reinforcement is simplified with an equivalent thickness d_{eq}. A local axis *xyz* is applied to the grid element, where the equivalent thickness in *x* axis and *y* axis are derived from longitudinal reinforcement and cross reinforcement respectively. Schematic diagram is as shown in Fig. 3, Where ϕ is the diameter of reinforcement in this local direction, *S* is the corresponding spacing of the reinforcement.

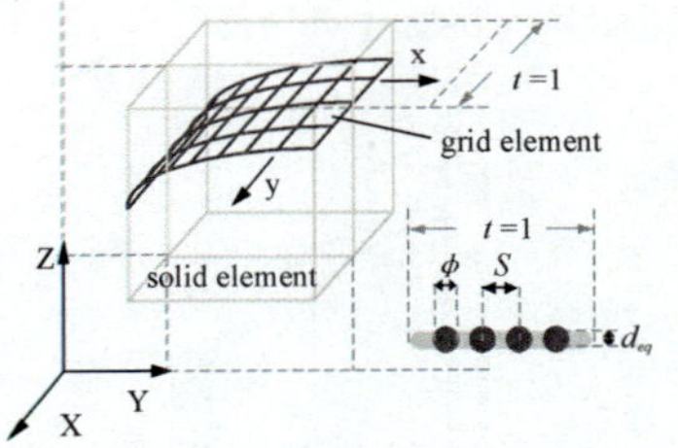

Fig. 3. Schematic diagram of grid reinforcement

Based on design drawing of the shield segment, the reinforcement is simplified into two layers of grid element inside and outside along the thickness direction. Von-Mises plastic strain model is endued to the reinforcement. The parameters of the reinforcement are shown in Table 2.

2.3 Soil Model

HS-small model (Benz 2007) is widely accepted in the numerical simulation of soil such as sand, silt and clay, having access to simulate the nonlinear behavior of soil loading and unloading. The Hyperbolic reduction curve of shear moludus is shown in Fig. 4, where the governing equation of tangent shear modulus is defined as a polyline:

$$G = \begin{cases} G_0\left(\dfrac{\gamma_{0.7}}{\gamma_{0.7} + \alpha\gamma_{Hist}}\right)^2 & \text{for } \gamma_s \leq \gamma_c \\ \dfrac{E_{ur}}{2(1 + v_{ur})} & \text{for } \gamma_s \geq \gamma_c \end{cases} \tag{7}$$

Table 2. Parameters of reinforcement

Property	Value
Young's modulus, E(MPa)	2×10^5
Poisson's ratio,v	0.3
Concrete cover thickness, c(mm)	60
Equivalent thickness, f_c(mm)	1.979 (outside)
	2.037 (inside)
Yielding strength, f_y(MPa)	335
Ultimate strength, f_u(MPa)	455

$$\gamma_c = \frac{\gamma_{0.7}}{0.385}\left(\sqrt{\frac{G_0}{G_{ur}}} - 1\right) \tag{8}$$

In which G_0 is small strain shear stiffness; $\gamma_{0.7}$ is the threshold shear strain when $G = 0.7G_0$; γ_c is the cut-off shear strain which is derived from Hardin-Drnevich relationship, as shown in Eq. (8); γ_{Hist} is the history strain; E_{ur} is the unloading-reloading stiffness; v_{ur} is the Poisson's ratio for unloading-reloading, which is equal to Poisson's ratio v by default and is a constant value.

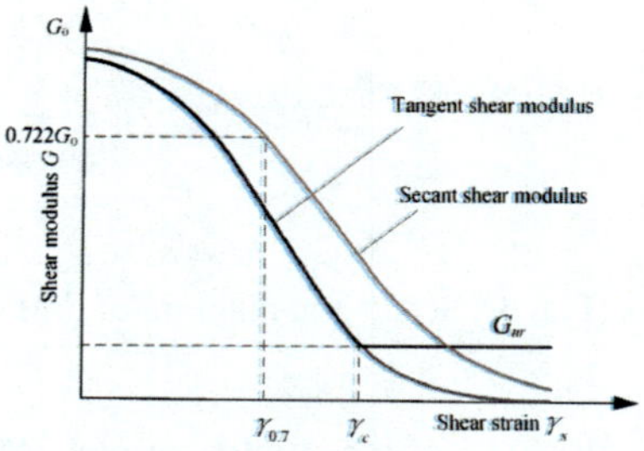

Fig. 4. Hyperbolic curve of shear modulus

The unloading-reloading stiffness E_{ur} and small strain shear stiffness G_0 are subject to depend on the current pressure according to the following power laws:

$$G_0 = G_0^{ref}\left(\frac{p_t' + p'}{p_{ref}'}\right)^m \tag{9}$$

$$E_{ur} = E_{ur}^{ref}\left(\frac{p_t' + p'}{p_{ref}'}\right)^m \tag{10}$$

In which G_0^{ref} is the reference small strain shear stiffness; E_{ur}^{ref} is the unloading-reloading shear stiffness; p_{ref}' is the reference pressure for stiffness; m is the power for stress-level dependency of stiffness (default $m = 0.5$); p_t' is a tensile pressure to ensure non-zero stiffness.

In order to facilitate the parameter analysis, the typical soil layer ③$_5$ in the real case is selected as a single layer. The parameters of the soil layer in HS-small model are shown in Table 3.

Table 3. Parameters of soil with HS-small model

Property	Value
Natural unit weight, ρ(kg/m^3)	1860
Poisson's ratio,v	0.2
Friction angle, ϕ(°)	6.5
cohesion, c(kPa)	29
Threshold shear strain, $\gamma_{0.7}(10^{-4})$	2
Reference pressure, p'_{ref} (kPa)	100
Reference small strain shear stiffness, G_0^{ref} (MPa)	43.8
Unloading-reloading shear stiffness, E_{ur}^{ref} (MPa)	24.3

2.4 Interaction Models

The interfaces included in the numerical model are structure-ground interface and longitudinal joint interface between segments. The structure-ground interface is simulated by Coulomb friction contact, and the friction angle of $\sin\varphi = 0.3$ is taken to illustrate the small sliding (Liu 2012).

Longitudinal joint is a three-dimensional complex contact where plenty materials need considering. Its contact behavior is closely related to packer, concrete and bolts. Thus, the simplified of longitudinal joint is essential in the layer-structure model. Herein the longitudinal joint is simplified as a two-dimensional 6-node line interface. It has been proved by comparison with full-scale test results that this simplified method can effectively illustrate the nonlinear behavior of longitudinal joint (Yang et al. 2016). The stiffness of interfaces is as shown in Table 4.

2.5 Mesh Strategies and Construction Phase

According to the covering soil of the real case, the surface loading is simplified to a trapezoidal load. CQ16E eight-node quadrilateral element is adopted in the mesh work, and different mesh sizes are set for soil and tunnel structure respectively. The size of the soil element is 1m, while the lining element is 0.05 m. Thus, the segment in thickness id divided into 6 elements which meets the requirements of nonlinear analysis for concrete structures. The mesh work of the model is shown in Fig. 5. The construction phase is performed with four steps. First step is the initial phase which balances the ground

Table 4. Interface stiffness

Property	Ground-structure	Longitudinal joint
Type	Coulomb friction	Linear elastic
Normal stiffness, k_n(N/m^3)	2×10^{12}	1×10^{12}
Shear stiffness, k_s(N/m^3)	1×10^{8}	3×10^{10}
Friction angle, φ(°)	17.45	-

stress. Then the linings and its interface are activated. In the third phase the soil in the tunnel is freezed to simulate tunnel excavation. After the tunnel excavation, the previous displacement is cleared, and the surface load is activated when the load factor is set to 0, 0.25, 0.5, 0.75 and 1.0 in turn.

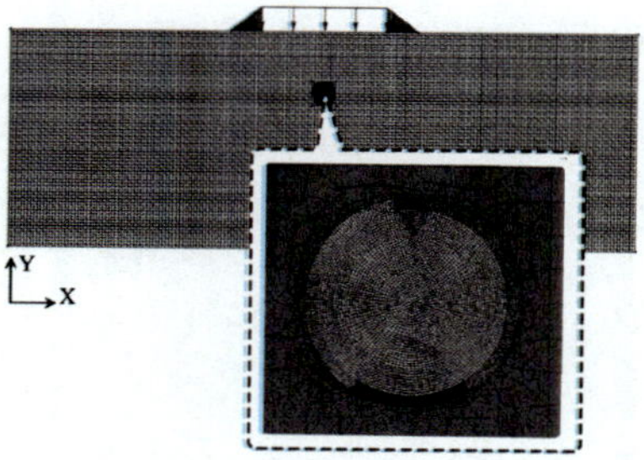

Fig. 5. FEM model and meshing

3 Parameter Analysis

Herein a sensitivity analysis is performed with several parameters of interest such as the rotation angle of the ring, loading width, the offset of loading and the depth of the tunnel. The schematic model is shown in Fig. 6, where x_1 and x_2 are the upper top width and the isosceles width of the trapezoid load respectively; q_0 represents the value of uniform load. Herein x_2 and q_0 are set to fixed value, which equal to 10 m and 140 kPa. Hence, the loading range can be defined by x_1. To analyze the influence of different assembly location for segments, an angle denoted by θ was used to define the relative rotation of the assembling lining. For example, when $\theta = 0$, the key segment is located at vault. Meanwhile, d and h denote the offset of loading and the depth of the tunnel. The parameters of cases are shown in Table 5.

In order to evaluate the damage state of the lining comprehensively, indicators of crack ratio (Yang et al. 2016) and crack index I_{cr} are introduced. They are used to evaluate the overall damage of the lining and predict the development of the cracks respectively, which are defined as:

$$\text{crack ratio} = \frac{\text{quantity of cracking integration points}}{\text{quantity of integration points of linings in total}} \times 100\% \tag{11}$$

Table 5. Case studies

Case	$\theta(°)$	d(m)	x_1(m)	h(m)
1	0	0	40	15
2	22.5	0	40	15
3	45	0	40	15
4	22.5	10	40	15
5	22.5	20	40	15
6	22.5	30	40	15
7	22.5	0	200	15
8	22.5	0	20	15
9	22.5	0	10	15
10	22.5	0	40	25
11	22.5	0	40	35
12	22.5	0	40	45

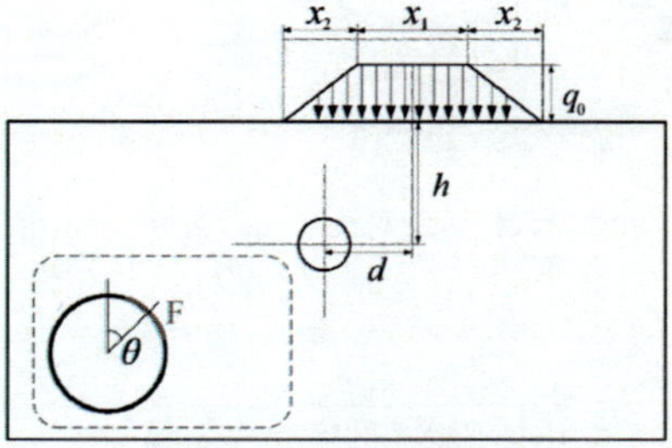

Fig. 6. Schematic model

$$I_{cr} = \frac{f_t}{\sigma_{\mathrm{I}}} \tag{12}$$

4 Analysis of Results

Figure 7 and 8 show the result of lining response of the cases given in the Table 4. The outer rings in Fig. 7 illustrate the contours of crack width, while the colored area of the contour indicates that the current area has reached the damage state. Meanwhile, it can be summed up that all the calculation cases show similar 4 crack area, which are the inner side of the vault and arch bottom, as well as outer side of the left and right hance. The inner rings in Fig. 7 illustrate the contours of I_{cr} with an absolute deformation of 50 times. According to the definition of Eq. (9), I_{cr} can be regarded as a safety margin of concrete. The closer its value is to 1, the more likely it cracks. Hence, the red area in contours is treated as "safe zone" where the tensile strength is much larger than the current principal stress, and the blue area is the "danger zone" that cracks have occurred or are about to appear. The influence of several parameters is set forth in subsequent sections respectively.

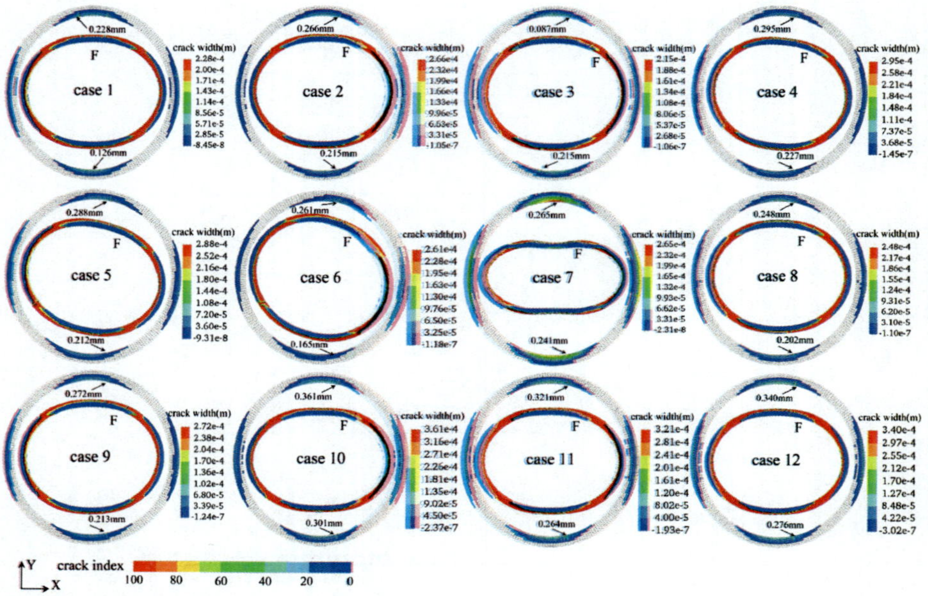

Fig. 7. Crack width and crack index contour (deformation enlarged 50 times) when load factor = 1

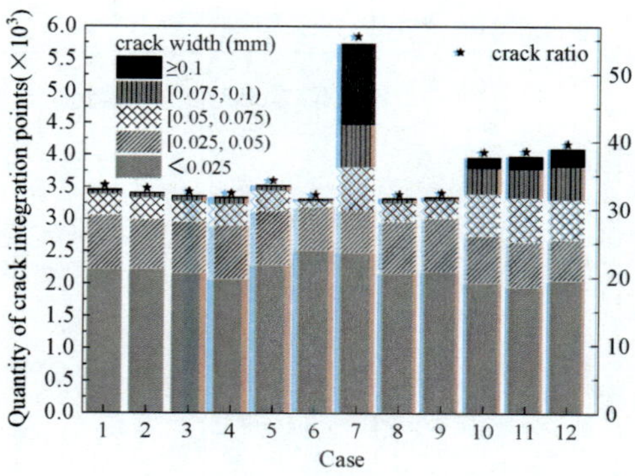

Fig. 8. Quantities of crack integration points when load factor = 1

4.1 Influence of Rotation Angle of the Ring

As aforementioned, the angle θ represents the overall rotation of the longitudinal joint position, thus it affects the force mechanism under loading of the tunnel at the same time. θ in cases 1–3 is set to 0°, 22.5° and 45° respectively.

It can be seen from Fig. 7 that when θ changes, the ovalisation and cracking area of the linings are basically unaltered, but the maximum crack width and its occurrence position changes due to the rotation of the longitudinal joint position. The occurrence position of maximum crack width often appears on one side of the longitudinal joint. However, it does not appear in the F segment, for the F segment is much shorter which leads to further stress release. When θ is 0°, the cracking contour is symmetrical; when it is 22.5°, the maximum crack width on the inner side of the vault and arch bottom increases; when it is 45°, there is no longitudinal joint in the cracked area inside the vault. Consequently, the maximum crack width of the vault decreases to 0.087 mm.

The I_{cr} contour shows that as θ increases, the distribution of the "danger zone" is basically unaltered, only slightly distinguished at each joint position. It can be seen from Fig. 8 that as θ increases, the cracking rate slightly decreases. When Load factor = 1, the cracking rates of three cases are 33.5%, 33.0%, and 32.5% respectively. To sum up, the increase of θ affect the internal force distribution of the ring due to joint shifting, and slightly improve the ability of crack resisting of the ring.

4.2 Influence of the Load Offsetting

Load offsetting is an important factor for the delimiting of the subway protection area, herein the influence of the load offsetting on the tunnel response is studied in cases 4–6. Figure 7 shows that when d increases, the four main crack areas rotate clockwise, while the I_{cr} contours also turn out with similar changes. When $d = 10$ m, the crack width on the vault and the arch bottom reaches maximum. As d increases, the maximum crack width decreases on the contrary.

Figure 9 show the change rule of newly added crack integration points when d increases from 0 to 100 m. When $d = 20$ m, the crack ratio reaches the maximum which equals to 34.2% when load factor = 1. when d is greater than 30 m, the crack ratio keeps dropping. During the offset of the surface loading, the horizontal displacement field where the tunnel is located changes, which means a forcible displacement in the opposite direction of d is imposed on the tunnel. Therefore, even if the load is not exactly symmetric with the tunnel, the tunnel may still have an even more unfavorable development trend of crack disease.

When the load factor = 0.25, incremental crack integration points decrease significantly when d increases. When d is greater than 50 m, the surface loading can barely cause an adverse effect of the lining. But in the subsequent load steps, the incremental crack integration points grow obviously when d approaches 100 m. Hence, the effect of load offsetting on the lining response still links to the magnitude of the surface loading. To sum up, the load offsetting will cause the overall rotation of the tunnel structure and crack area. Within a certain offsetting range, it may bring more adverse effects such as the increase of the maximum crack width and crack ratio. But when d is greater than 30 m in the case, the crack ratio keeps dropping as d increases.

4.3 Influence of Load Width and Buried Depth of Tunnel

The parameter analysis of cases 7–9 and cases 10–12 are carried out for x_1 and h respectively. Case 7 sets x_1 to 200 m, its crack ratio reaches to 55%, which is much higher than

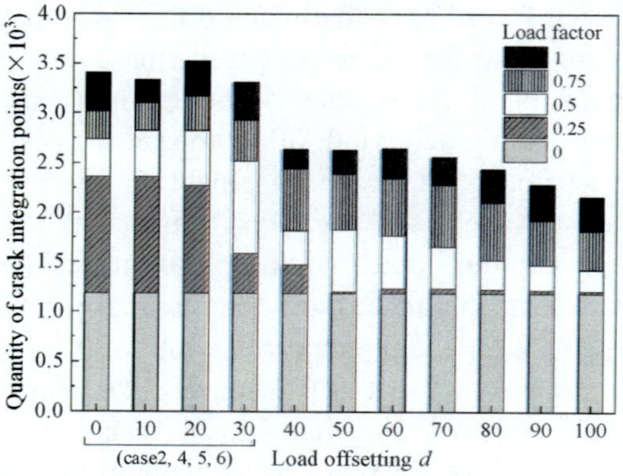

Fig. 9. Quantities of incremental crack integration points in each load case

any other cases, and the crack area has been extended to the full ring. Case 8 and Case 9 reduce x_1 to 20 m and 10 m, whose crack ratios are 32.2% and 32.4% respectively.

h in cases 10–12 are 25 m, 35 m and 45 m respectively, whose crack ratio are 38.4%, 38.5% and 39.6%. The reason why it leads to a slight increase on the crack ratio as h increases can be explained as follows. Herein, the surrounding soil is treated as soft soil, while the stress redistribution after the tunnel excavation is considered. Therefore, as buried depth increases, the stress condition of the tunnel deteriorates before surface loading. As a result, the increase in buried depth under surface loading in soft soil area may not bring beneficial effects.

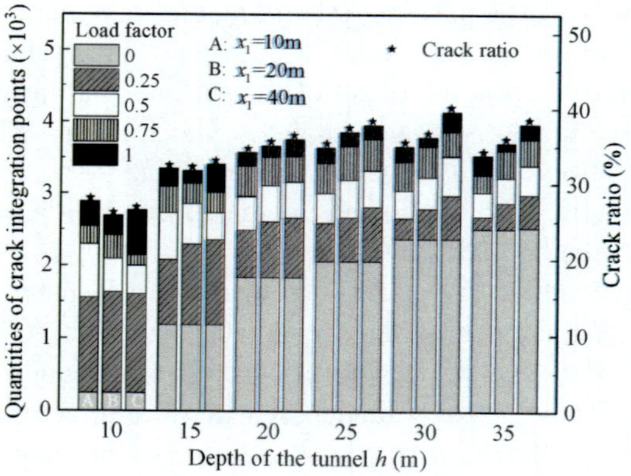

Fig. 10. Effect on crack state of load width and tunnel depth

Considering that the width of the surface loading effects the depth of the additional stress field, it is inappropriate to separate load width from the buried depth of the tunnel in the parameter analysis. Therefore, supplementary cases of double-factor parameter analysis are carried out, as illustrated in Fig. 10. When h is less than 15 m, the crack ratio under the 10 m-width surface loading is the highest, followed by 40 m and 20 m;

when h is equal to 15 m, the crack ratio of three load widths is basically the same; when h continues to increase, the larger the loading width is, the higher the crack ratio will be.

From the perspective of the effect of buried depth on crack ratio, as h increases, the effect of the surface load on the crack ratio decreases. However, if the adverse effect of stress redistribution after the tunnel excavation is considered, as h increases, the crack ratio will increase first and then decrease, while the peak of crack ratio is regarded as the most unfavorable buried depth. $x_1 = 10$ m corresponds to the most unfavorable burial depth about 25 m, with a crack ratio of 35.3%; $x_1 = 20$ m corresponds to the most unfavorable burial depth about 25 m, with a crack ratio of 37.4%; $x_1 = 40$ m corresponds to the most unfavorable burial depth about 30 m, with a crack ratio of 40.2%.

4.4 Discussion on the Relationship Between Ovalisation and Crack Ratio

Several parameters discussed above can affect the tunnel crack state to varying degrees. However, they cannot be separated to establish a direct relationship with current crack state. The crack rate is an evaluation index for lining damage state which is difficult to obtained by detection in reality, hence it is vital to establish a connection between the crack ratio and commonly used monitoring index.

Tunnel ovalisation, as a commonly used structural evaluation index, has been proved to evaluate the tunnel structural performance effectively (Shao et al. 2016; Wang and Zhang 2013; Zhou et al. 2019). Based on the calculation results in Sect. 4.3, the fitting curve of the relationship between the horizontal ovalisation and crack ratio is carried out, as shown in Fig. 11. The results show that the power function curve fits well with the scatter plot, revealing the nonlinear relationship between crack ratio and the horizontal ovalisation: as horizontal ovalisation increases, the crack ratio increases but the its growth rate slows down. Therefore, the power function can be used as a simple method to evaluate the damage state of tunnel ring.

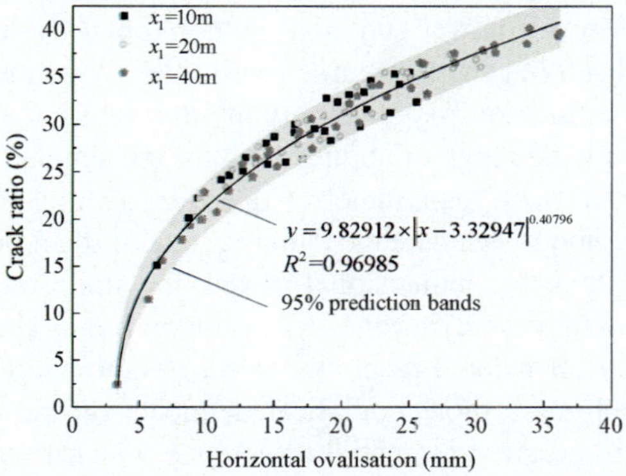

Fig. 11. Relationship curve of horizontal ovalisation and crack ratio

5 Conclusions

Based on the layer structure method, a two-dimensional numerical modelling for segmental lining subjected to surface loading involving the nonlinear behavior of both soil and lining is carried out. Thereby a direct connection between surface loading, soil and tunnel is performed, which realizes a one-step numerical simulation from surface loading to structural response. The following conclusions can be drawn through the parameter study:

- Maximum crack width, crack index, and crack ratio are introduced to evaluate the damage characteristics of linings. Maximum crack width points out the most detrimental area where diseases are most likely to occur in practice. The crack index evaluates the tensile safety margin of structure, thereby predicting the trend of further cracking. The crack ratio clarifies the percentage of cracked integration points, which can be regarded as an important indicator to evaluate the overall damage state of the ring.
- The increase of rotation angle of the ring affects the internal force distribution of the ring, and slightly improves the crack-resistant ability of the ring. Within a certain offset range, the offset of the non-axisymmetric surface loading may bring consequences like the overall rotation of the damage zone, growth in the crack ratio and increasing in the maximum crack width. But when d is greater than 30 m in the case, the crack ratio keeps dropping as d increases.
- Double-factor parameter analysis for the width of the surface loading and the buried depth of the tunnel shows that when the tunnel is buried shallow, a surface loading whose width is too small or too large will lead to an increase in crack ratio. When it is buried deeper, a larger load width has a greater impact on crack response. If the stress redistribution after tunnel excavation is not taken into consideration, the increase in buried depth can effectively reduce the adverse effects of surface loads. However, if the adverse effects of surface load and stress redistribution after excavation are considered at the same time, there is an unfavorable buried depth that maximizes the crack ratio of the tunnel.
- The fitting curve of the relationship between the crack ratio and the horizontal ovalisation performs a power function: as horizontal ovalisation increases, crack ratio increases rapidly at the beginning, while the growth rate continues to slow down. However, the function's applicability in other types of shield tunnel needs verifying. Still, it has a wide range of applications for the standard ring in the case is universal in metro tunnels and road tunnels. Consequently, this power function can be used as a simple method to evaluate the damage state of the tunnel ring.
- Carried out in a two-dimensional model, the study only focuses on the transverse ovalisation of the tunnel ring and its longitudinal crack characteristics, when the impact of the three-dimensional assembly effect is ignored. The longitudinal deformation of the tunnel has a more complicated impact on the crack behaviors both in the longitudinal and circumferential directions, which needs figuring out in the future work.

References

Shao, H., Huang, H.W., Zhang, D.M., Wang, R.L.: Case study on repair work for excessively deformed shield tunnel under accidental surface surcharge in soft clay. J. Chin. J. Geotech. Eng. **38**(6), 1036–1043 (2016)

Arnau, O., Molins, C.: Three dimensional structural response of segmental tunnel linings. J. Eng. Struct.. **44**, 210–221 (2012)

Zhou, B., Xie, X.Y., Yang, Y.B., Wang, X.J.: Service performance evaluation and cross scale simulation of shield tunnel during overhaul. J. Tongji Univ. (Nat. Sci.) **47**(10), 1390–1397 (2019)

Chen, J.S., Mo, H.H.: Numerical study on crack problems in segments of shield tunnel using finite element method. J. Tunnel. Undergr. Space Technol. **24**(1), 91–102 (2009)

Wang, R.L., Zhang, D.M.: Mechanism of transverse deformation and assessment index for shield tunnels in soft clay under surface surcharge. J. Chin. J. Geotech. Eng. **35**(6), 1092–1101 (2013)

Papanikolaou, V.K., Kappos, A.J.: Practical nonlinear analysis of unreinforced concrete tunnel linings. J. Tunnel. Undergr. Space Technol. **40**, 127–140 (2014)

Xu, G., He, C., Lu, D., Wang, S.: The influence of longitudinal crack on mechanical behavior of shield tunnel lining in soft-hard composite strata. J. Thin. Walled Struct. **144**, 106282 (2019)

Sun, L.W., Qin, J.S., Hong, Y., Wang, L.Z., Zhao, C.J., Qin, X.: Shield tunnel segment and circumferential joint performance under surface surcharge. J. ZheJiang Univ. Eng. Sci. **51**(8), 1509–1518 (2017)

Yang, Y.B., Zhou, B., Xie, X.Y.: Study on transverse deformation and cracking property of shield-driven tunnel induced by adjacent excavation. J. Chin. J. Rock Mech. Eng. **35**, 4082–4093 (2016)

Xie, J.C., Wang, J.C., Huang, W.M.: Nonlinear structural analysis on cracking behavior of TBM tunnel segment under surface loading. J. Railway Sci. Eng. in press (2020)

Hendriks, M.A.N., de Boer, A., Belletti, B.: Guidelines for nonlinear finite element analysis of concrete structures. J. Rijkswaterstaat Technisch Document (RTD), Rijkswaterstaat Centre for Infrastructure, RTD **1016** (2017)

Feenstra, P.H.: Computational aspects of biaxial stress in plain and reinforced concrete. PhD thesis, Delft University of Technology, Delft, The Netherlands (1993)

Benz, T.: Small-strain stiffness of soils and its numerical consequences. PhD thesis, University of Stuttgart, Institute of Geotechnik, Stuttgart, Germany (2007)

Vecchio, F.J., Collins, M.P.: Compression response of cracked reinforced concrete. J. Struct. Eng. **119**(12), 3590–3610 (1993)

Liu, H.: Three-dimensional analysis of underground tunnels in liquefiable soil subject to earthquake loading. In: GeoCongress 2012: State of the Art and Practice in Geotechnical Engineering, pp. 1819–1828 (2012)

The Freezing Characteristic Curve of a Coarse-Grained Volcanic Soil

Junping Ren[1(✉)], Shoulong Zhang[2], Tatsuya Ishikawa[2], and Sai K. Vanapalli[3]

[1] College of Civil Engineering and Mechanics, Lanzhou University, 222 South Tianshui Road, Lanzhou 730000, China
renjp@lzu.edu.cn

[2] Division of Field Engineering for the Environment, Hokkaido University, Kita 13, Nishi 8, Kita-Ku, Sapporo 060-8628, Japan

[3] Department of Civil Engineering, University of Ottawa, 161 Louis-Pasteur Street, Ottawa K1N6N5, Canada

Abstract. Volcanic soils are widely distributed in the Hokkaido prefecture of Japan, where seasonal freezing and thawing occur. In frozen soils, a portion of pore water remains unfrozen, due to the effects of capillarity, adsorption, and possibly solute. The variation of the amount of unfrozen water in a frozen soil, which is primarily influenced by subzero temperature, has great impacts on the physical and mechanical behavior of the soil. In the present study, the soil-freezing characteristic curve (SFCC) of a typical volcanic soil sampled in Hokkaido, namely, the Komaoka soil, was investigated. The unfrozen water content of the prepared Komaoka soil specimens was measured using a cheap and convenient moisture sensor. The temperature of the specimens was determined by a rugged temperature sensor. Different number of freeze-thaw (F-T) cycles, and different freezing/thawing methods (i.e. one- and three-dimensional) were considered, and their effects on the SFCC were investigated. The experimental results suggest that neither of the F-T cycles nor the freezing/thawing methods had significant influence on the measured SFCC. The present study can contribute to understanding the freezing characteristic of the investigated soil in specific and that of other cold region soils in general.

Keywords: Frozen soils · Unfrozen water · Soil-freezing characteristic curve · Subzero temperature · Freeze-thaw cycles

1 Introduction

The unfrozen water and pore ice coexist within a frozen soil. The unfrozen water exists in small pore spaces and as thin films adsorbed on the surfaces of soil particles in equilibrium with the pore ice at subzero temperatures. The soil-freezing characteristic curve (SFCC) describes the relationship between the amount of unfrozen water and its energy state or subzero temperature in a frozen soil (Koopmans and Miller 1966; Spaans and Baker 1996; Azmatch et al. 2012; Watanabe and Osada 2016; Schafer 2018). This relationship has also been referred to as the phase composition curve (PCC) (Anderson and Tice 1972; Liu and Yu 2014), unfrozen water characteristic curve (UWCC) (Mu et al. 2018), unfrozen water content curve (Kozlowski 2003), soil freezing curve (SFC)

Y. Liu et al. (Eds.): GeoChina 2021, SUCI, pp. 80–96, 2021.
https://doi.org/10.1007/978-3-030-80316-2_9

(Grant and Sletten 2002), freezing characteristic curve (Smith 1985), and soil freezing retention curve (SFRC) (Mao et al. 2018).

The SFCC links the degree of phase transition to the subzero temperature in a frozen soil. Therefore, many important physical and mechanical properties of frozen soils can be estimated using SFCC. In addition, the constitutive relationships for hydraulic, thermal, and mechanical fields of frozen soils are functions of the quantity of unfrozen water, the SFCC is therefore essential to modeling the transport mechanism of water, heat, and solutes in frozen soils (Spaans and Baker 1996; Zhang et al. 2016; Yu et al. 2018; Xu et al. 2020).

The SFCC of various coarse- and fine-grained soils have been investigated in the literature. The present study focuses on the experimental determination of the SFCC of a volcanic soil. Volcanic soils cover 1% of the Earth's surface (or more than 124 million hectares) yet support 10% of the world's population, including some of the highest human population densities (Neall 2009). One of the major areas of volcanic soils rim the Pacific and occur in countries such as Japan, Philippines, United States, Ecuador, and Peru. In the Hokkaido of Japan, there are over 40 Quaternary volcanoes, and pyroclastic materials cover over 40% of these areas. Significant volcanic activity occurred in the Neogene's Quaternary period, and various pyroclastic materials, such as volcanic ash, pumice and scoria, formed during those eruptions (Kawamura and Miura 2013). The volcanic soils show different behavior from that of clay or sand, and have been classified as problematic soils in 1980s, according to IS-Tohoku98 (Kawamura and Miura 2013). The Komaoka soil, which is a typical volcanic soil found in Hokkaido, is investigated in the present study. The original soil sample was sampled from the ejectas of the Shikotsu calderas near the Sapporo city, as shown in Fig. 1.

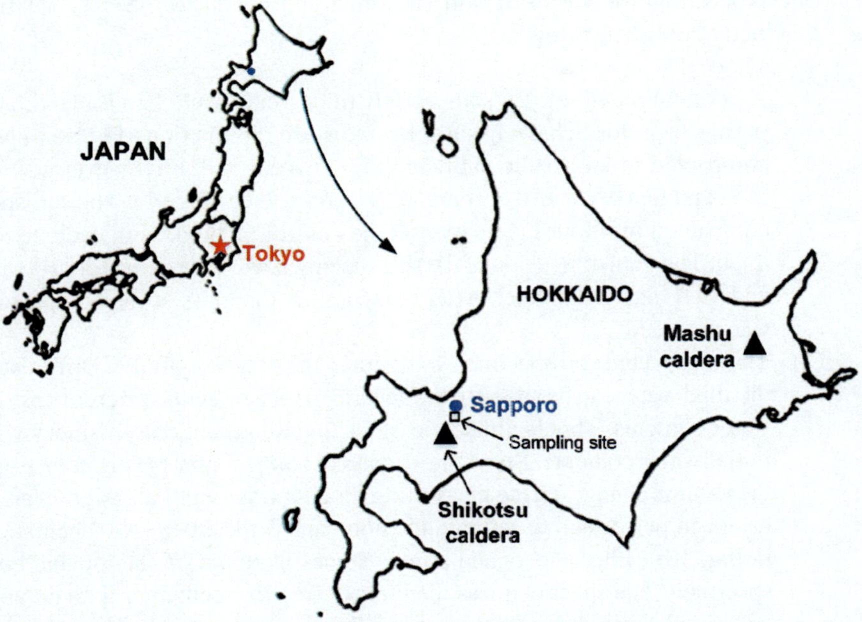

Fig. 1. Location of the sampling site

2 Experimental Setup for SFCC Measurement

The Komaoka soil contains a considerable amount of highly porous particles. These highly porous particles are weak and easy to break by hands. Many pores can be seen by naked eyes. This largely contributes to the low bulk density of the Komaoka soil. For example, the in-situ dry density of the Komaoka soil is around 0.794 g/cm^3 (Nguyen 2017). In the present study, the soil particles with sizes smaller than 2 mm were used for the SFCC test. The basic properties of the Komaoka soil is summarized in Table 1, and its gradation curve is shown in Fig. 2. The Komaoka soil has a very low fraction of clay and is a non-plastic coarse-grained soil. For comparison, the gradation curve of the Japanese standard sand, Toyoura sand, is shown in Fig. 2. The Toyoura sand has a specific gravity of 2.65 and zero percentage of fines (Ishikawa et al. 2013).

Table 1. Physical properties of the Komaoka soil

G_s	ρ_{dmax} (g/cm^3)	ρ_{dmin} (g/cm^3)	w_0 (%)	ρ_d (g/cm^3)	w_{hygro} (%)	% Sand	% Silt	% Clay	C_u	C_c
2.50	1.12	0.76	≈30	0.915	0.96	64	28	8	45	1.54

Note: G_s: specific gravity; ρ_{dmax} and ρ_{dmin}: the maximum and minimum dry density, respectively; w_0: the natural gravimetric water content; ρ_d: dry density; w_{hygro}: hygroscopic water content; C_u: coefficient of uniformity; C_c: coefficient of curvature. Values for G_s, ρ_{dmax} and ρ_{dmin} are from Kawamura and Miura (2013); value for w_0 is from Nguyen (2017)

2.1 Experimental Setup

The procedures for preparing soil specimen and conducting SFCC measurement are succinctly described below.

(1) A certain amount of dry soils and distilled water were hand mixed, and stored in plastic bags for achieving uniform moisture distribution. The soil specimen was compacted in an acrylic cylinder (Fig. 3(a)), at the gravimetric water content of 30% and to a dry density around 0.915 g/cm^3 (see Table 1). The soil specimen was compacted into four layers by volume control method, with each layer of 25 mm thick. The final dimension of the soil specimen was approximately 100 mm in height and 49.6 mm in diameter. After compaction, the mass of the compacted specimen was recorded.

(2) The compacted soil specimen was kept in the acrylic cylinder. Different amounts of distilled water can be added to the specimen for achieving different post-compaction water contents. This facilitate the SFCC measurement on specimens with different initial water contents. Since the Komaoka soil is a non-plastic coarse-grained soil, the volume change during the wetting process was assumed insignificant. A trial test has been performed to saturate the compacted specimen. As schematically shown in Fig. 3(b), filter papers and porous stones were put on the top and bottom of the specimen. The specimen was then transferred to a container with de-aired distilled water and subjected to vacuum for full saturation for 12 h. However, it was found that the specimen collapsed after saturation, due to its weak water retention ability.

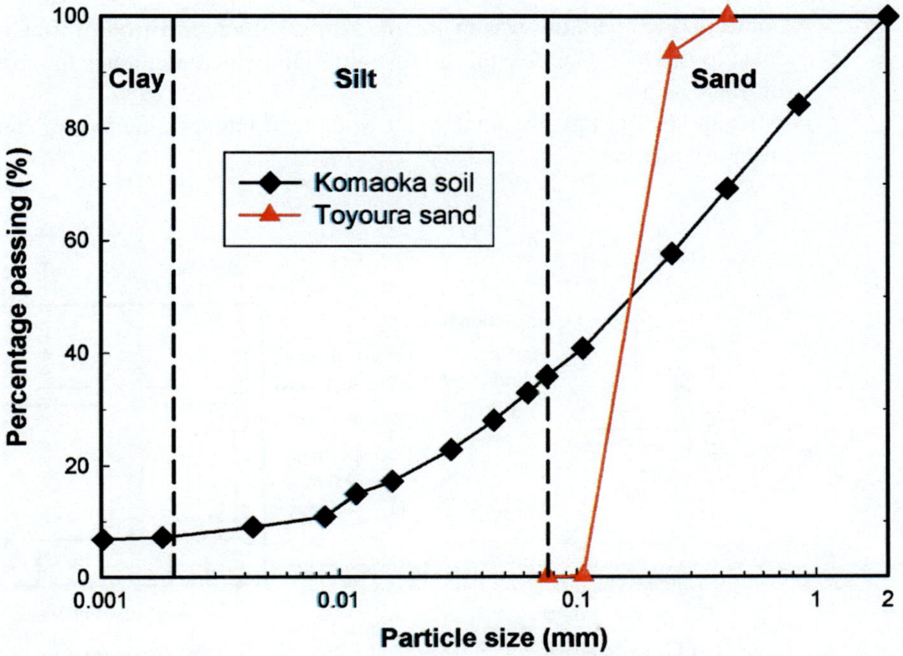

Fig. 2. Particle size distribution of the Komaoka soil and Toyoura sand (modified after Ishikawa et al. (2013) and Nguyen (2017))

(3) The prepared soil specimen was sealed by plastic bag for waterproof (as the specimen will be submerged in antifreeze liquid). One EC-5 moisture sensor (METER Group, Inc. USA) and one RT-1 temperature sensor (METER Group) were inserted into the specimen, and had good contact with the specimen. The two sensors were connected to an EM50 datalogger (METER Group). The time interval for water content and temperature measurements was generally 15 min. Figure 4(a) is an example showing the prepared specimens ready for SFCC test. The EC-5 moisture sensor and RT-1 temperature sensor were calibrated prior to their use, as summarized in the following section. It is assumed that the EC-5 sensor measures the volumetric water content of the bulk specimen. In other words, the sampling volume of the EC-5 sensor is equal to (or within) the volume of the specimen. Therefore, the effect of the surrounding antifreeze liquid on EC-5 reading is insignificant.

(4) The specimen was then submerged in antifreeze liquid inside a low-temperature bath for SFCC measurement, as shown in Fig. 3(c) and Fig. 4(b). The temperature of the antifreeze liquid can be controlled and different temperature paths (e.g., freezing and thawing) were exerted on the soil specimen for determining SFCC. Each

controlled temperature was maintained for a sufficient period of time to approximate to thermodynamic equilibrium condition, which facilitates the measurement of SFCC.

(5) After the SFCC test, the total water content of the specimen was determined by oven dry method.

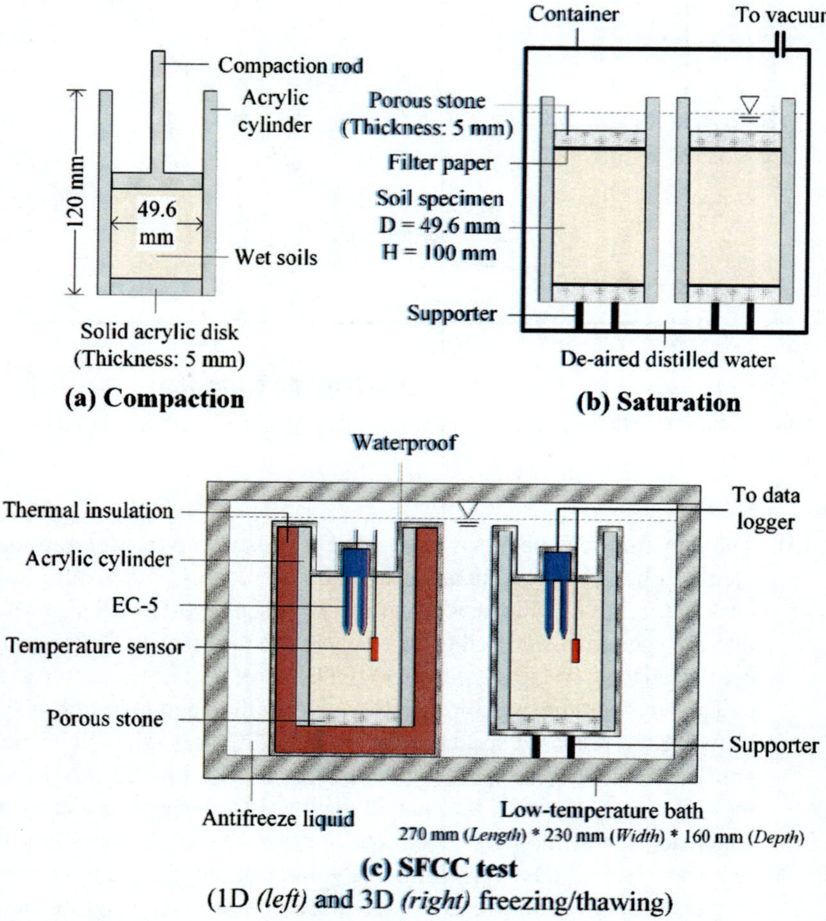

Fig. 3. Schematic diagram of specimen preparation and setup for SFCC test

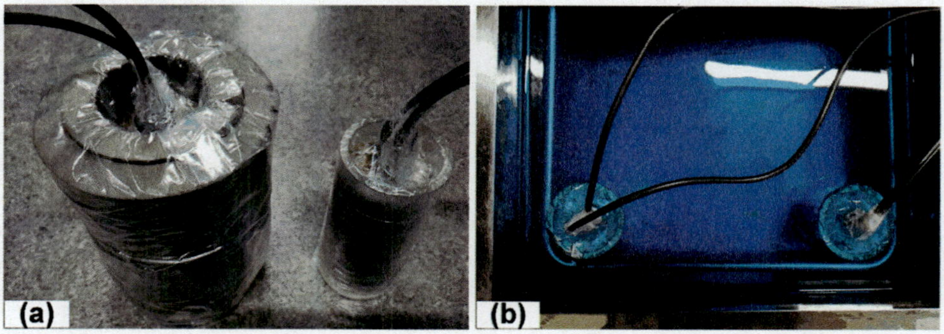

Fig. 4. (a) Soil specimens ready for SFCC test, and (b) specimens submerged in antifreeze liquid for measurement

2.2 Sensor Calibration

2.2.1 Temperature Sensor Calibration

The calibration for the RT-1 temperature sensors was fulfilled by referring their measured temperature values to those measured by a thermometer. A couple of different temperature values were selected. The result is shown in Fig. 5. It can be seen that the measured values by the temperature sensors are closely located on or near the 1:1 line. Therefore, it is assumed that the temperature sensors are accurate.

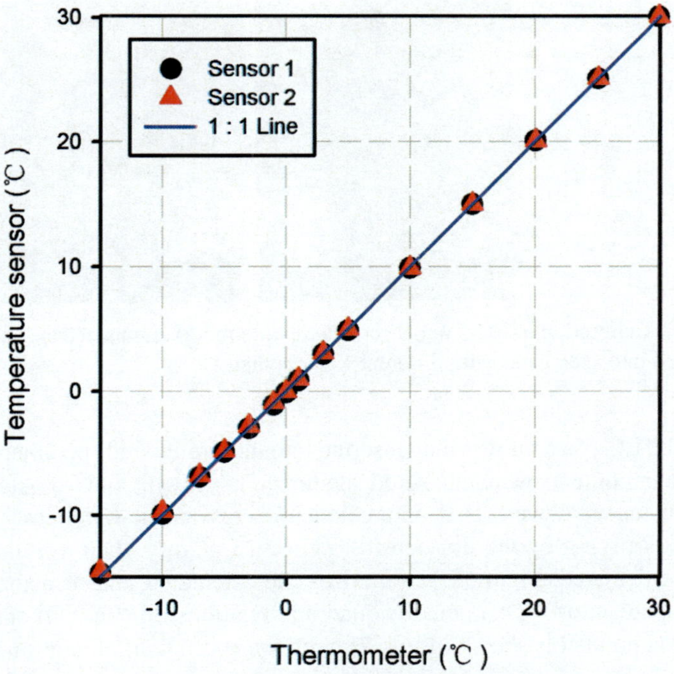

Fig. 5. Calibration of RT-1 temperature sensor

2.2.2 EC-5 Calibration Under Unfrozen Condition

For the reliable measurement of volumetric (unfrozen) water content, the specific calibration of the EC-5 moisture sensor for the Komaoka soil is required. In the present study, the calibration of the EC-5 sensor under room temperature was carried out. More than 20 specimens were prepared for this purpose. The specimens were compacted at the gravimetric water content of 30% and to a dry density around 0.915 g/cm^3 (by following the same procedures described above), as shown in Fig. 6(a). After compaction, several specimens were subjected to drying or wetting, by leaving the top/bottom surfaces of the specimen open to air or by adding distilled water to the specimen. The masses of the specimens were monitored during the drying and wetting processes, from which the desired water contents can be approached. As a result, specimens with different water contents were prepared.

Fig. 6. Calibration of EC-5 under room temperature, (a) compacted specimens, (b) and (c) EC-5 inserted into specimens with different water contents

The EC-5 sensor was then inserted into those prepared specimens and their volumetric water contents were measured, as shown in Fig. 6(b) and (c). After measurement, the specimens were oven dried, and their actual gravimetric water contents can be obtained. By multiplying by the dry density (i.e. 0.915 g/cm^3), their volumetric water contents were calculated. Figure 7 summarizes the calculated and measured volumetric water contents of those specimens. A quadratic relationship (Eq. (1)) can be used to best-fit the data points (as shown in Fig. 7), with the coefficient of determination of 0.98.

$$y = -1.49x^2 + 1.69x + 0.034 \tag{1}$$

where, y is the actual (calculated) volumetric water content, and x is the measured volumetric water content by EC-5.

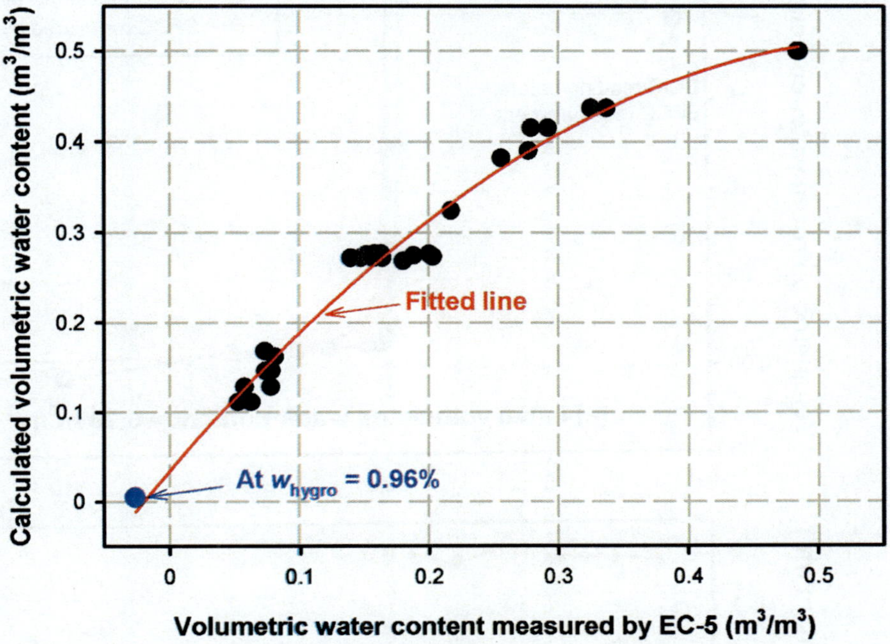

Fig. 7. Calibration of EC-5 sensor for the Komaoka soil under unfrozen condition

2.2.3 Accuracy of EC-5 Under Frozen Condition

A lack of reliable equipment prevents the calibration of the EC-5 sensor under frozen condition. In the present study, the accuracy of the EC-5 sensor at sub-freezing temperatures was investigated by comparing the EC-5 measured SFCC with that measured by NMR (nuclear magnetic resonance) on the Toyoura sand.

Watanabe and Wake (2009) measured the SFCC of Toyoura sand by using NMR. The Toyoura sand in their study has a mean particle diameter of 0.21 mm and a uniformity coefficient of 1.44. Although these two values are slightly different from those of the Toyoura sand employed in the present study, the effect of this difference on the measured SFCC is assumed negligible. As in Watanabe and Wake (2009), two Toyoura sand specimens were prepared, by mixing certain amount of dry sand and distilled water and by subsequent compaction in the acrylic cylinder shown in Fig. 3. The two specimens had different initial volumetric water contents (0.26 and 0.17 m^3/m^3), but the same bulk density (1.43 g/cm^3) and solid fraction (0.54 m^3/m^3). The Toyoura sand was previously washed in distilled water to minimize the effect of salinity, if any, on the SFCC. Following similar steps described in Sect. 2.1, the SFCC of the two Toyoura sand specimens were measured by using EC-5, along thawing path. The results are summarized in Fig. 8, together with the SFCC measured by NMR by Watanabe and Wake (2009).

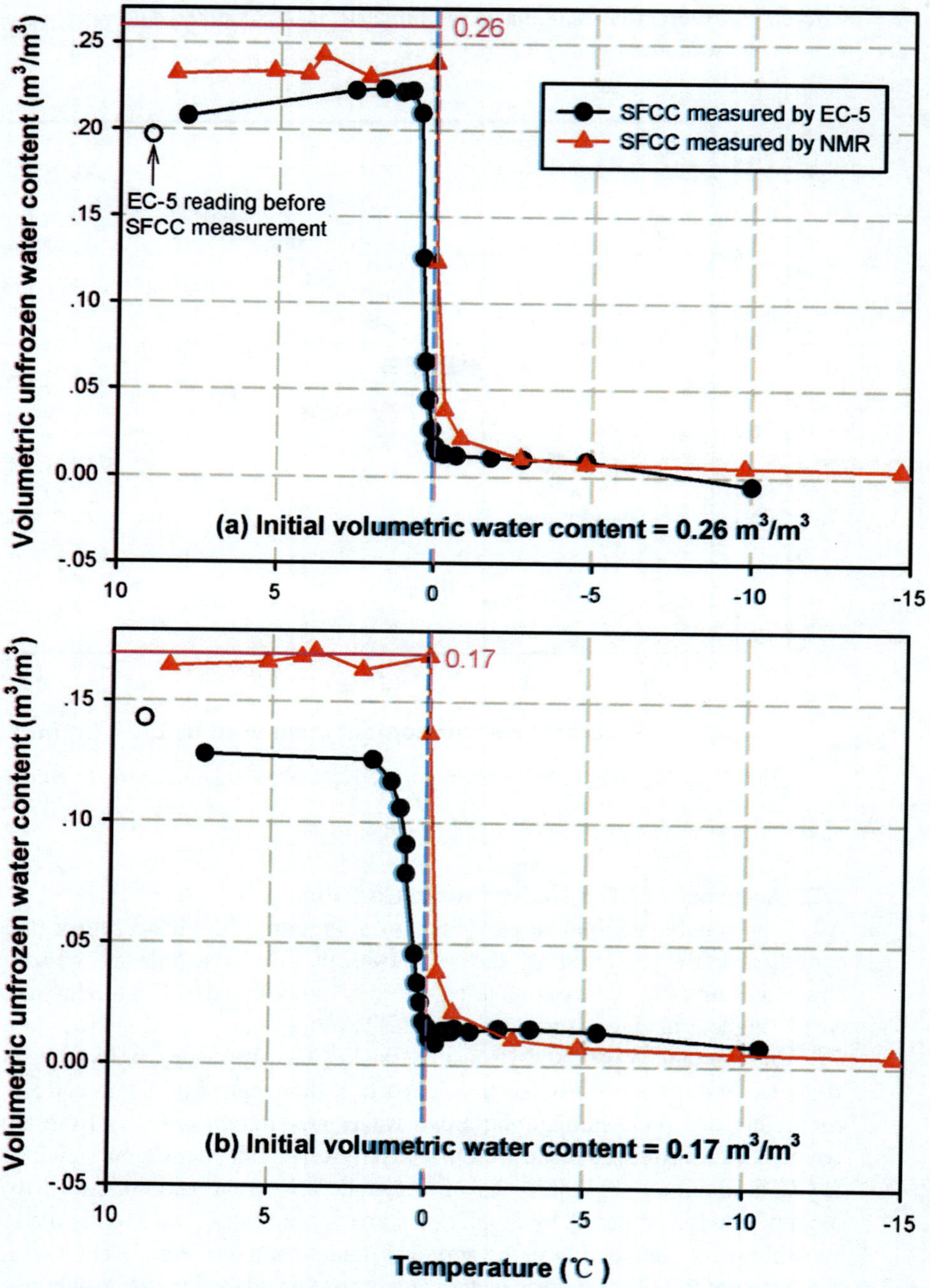

Fig. 8. Comparison of the measured SFCCs by EC-5 and by NMR for two Toyoura sand specimens with different initial volumetric water contents, (a) 0.26 and (b) 0.17

It can be seen that the SFCC measured by EC-5 shows similar shape or desorption rate to that measured by NMR. In other words, the EC-5 correctly captures the tendency of change in unfrozen water content of the soil specimens. However, the two SFCCs are not completely overlapped with each other. This means that there is difference between the volumetric water content values measured by EC-5 and NMR. Three temperature ranges can be distinguished,

(1) When the temperature of the specimen is relatively high (e.g., higher than 1 °C), the measured values by EC-5 and NMR are relatively stable. In this temperature range, the specimen is completely thawed or unfrozen, and the measured value represents the initial (total) water content of the specimen. The EC-5 sensor is able to give reasonable measurement in this range. However, specific calibration for the Toyoura sand can improve the accuracy of the EC-5 sensor in this temperature range.
(2) When the temperature of the specimen is relatively low (e.g., lower than −1 °C), the values measured by EC-5 and NMR show agreement. In this temperature range, most pore water in the specimen become ice and the unfrozen water content values are very low. The EC-5 sensor gives reasonable measurement in this range. And,
(3) When the temperature approaches 0 °C along the thawing process (e.g., between −1 and 1 °C), the difference between the SFCCs measured by EC-5 and NMR is large, even though their desorption rates are similar.

It is noted from the experimental data that in the first two temperature ranges, i.e. the specimen is either completely thawed or "completely frozen", the volumetric water content values measured by the EC-5 sensor is stable. In other words, when the measured temperature of the specimen kept constant during a sufficient period of time (which is typically taken as thermodynamic equilibrium is achieved), the measured volumetric water content maintained at a specific value during this time period. Within the third temperature range, however, the measured value by EC-5 continuously increased even though the measured temperature of the specimen kept constant. This may imply that pore ice underwent continuous phase change to water, and thermodynamic equilibrium was not actually achieved under the controlled temperatures.

The EC-5 sensor measures the volumetric water content of the bulk specimen. However, the temperature sensor measures the temperature of soil body that is directly contacted with it. Therefore, if thermodynamic equilibrium is not fully established, the measured temperature only represents the temperature of a portion of soil, rather than the temperature of the bulk specimen (Ren and Vanapalli 2019). In other words, the soil mass or volume sampled by these two types of sensors are not equal. The SFCC measured by EC-5 goes to the left side of the 0 °C line. The reason may be that the soil body that has direct contact with the temperature sensor is thawed and have a positive temperature, yet there is still frozen part present in the bulk specimen.

In summary, the measured SFCC by EC-5 sensor shows good result when no ice is present or most pore water changes phase to ice in the soil specimen. However, the measurements in the temperature range where massive phase change occurs may bear errors. Decrease the size of the specimen can facilitate thermodynamic equilibrium, as was assumed for the small NMR specimen (whose volume is only 3% of that of the specimen in the present study) in Watanabe and Wake (2009). However, this method may not be applicable to the EC-5 sensor because it requires a certain volume of soil to obtain a representative measurement. In addition, one difficulty in measuring SFCC of coarse-grained soils is that the phase change of most pore water occurs within a very narrow range of temperature. For example, this temperature range is 0 to −0.3 °C for the Toyoura sand specimens where the unfrozen water content drops from 0.24 or 0.17 to 0.04 m^3/m^3. The accuracy of controlling the specimen's temperature and the accuracy and resolution of typical temperature sensors are not good enough for measuring the SFCC of coarse-grained soils in this narrow temperature range.

3 Experimental Results and Discussion

Several Komaoka specimens have been used to conduct the SFCC measurement. Figure 9 shows the temperature-time curves for two specimens, which were subjected to three consecutive freeze-thaw (F-T) cycles. The specimens, which were initially under unfrozen condition, were directly subjected to −10 °C in each F-T cycle. The controlled temperature was then gradually increased to desired subzero temperatures for measuring the thawing branch of SFCC. It can be seen that supercooling was occasionally observed under the testing condition in the present study. The specimens have relatively high freezing temperature (i.e. around −0.2 °C), at which most pore water starts to change phase to ice with a fast rate. This is reasonable since the Komaoka soil is a coarse-grained soil and contains only a small fraction of clayey particles.

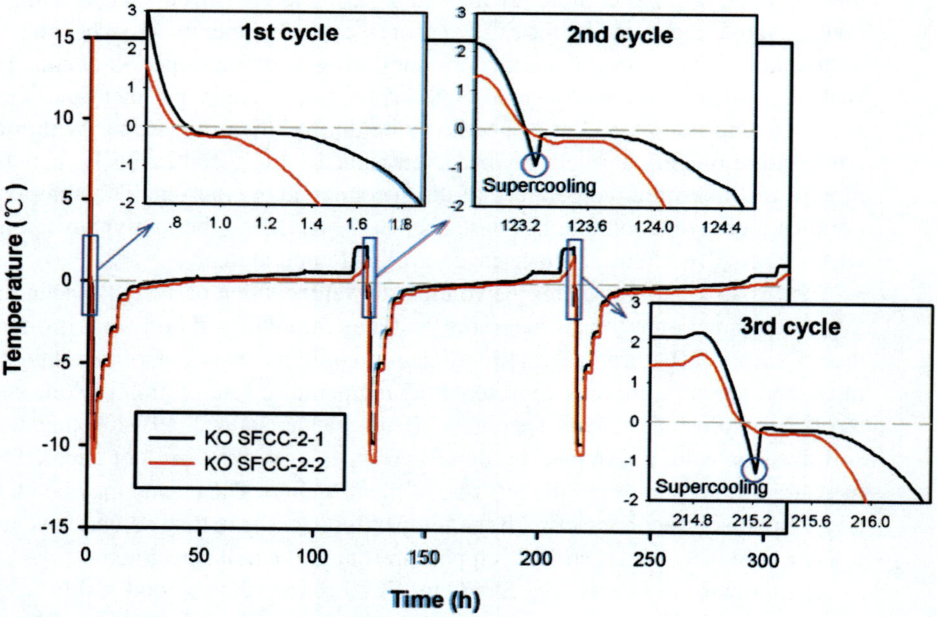

Fig. 9. The temperature-time curves of two Komaoka specimens subjected to three F-T cycles

Figure 10 summarizes the measured SFCCs of three Komaoka specimens. The three specimens had an initial gravimetric water content of 30.3%, 30.3% and 41.7%, respectively. By multiplying by the dry density of the specimens (i.e. 0.915 g/cm^3), their initial volumetric water contents are obtained and shown in the figure with pink lines and numbers. The EC-5 measured water content of the specimen at completely thawed condition

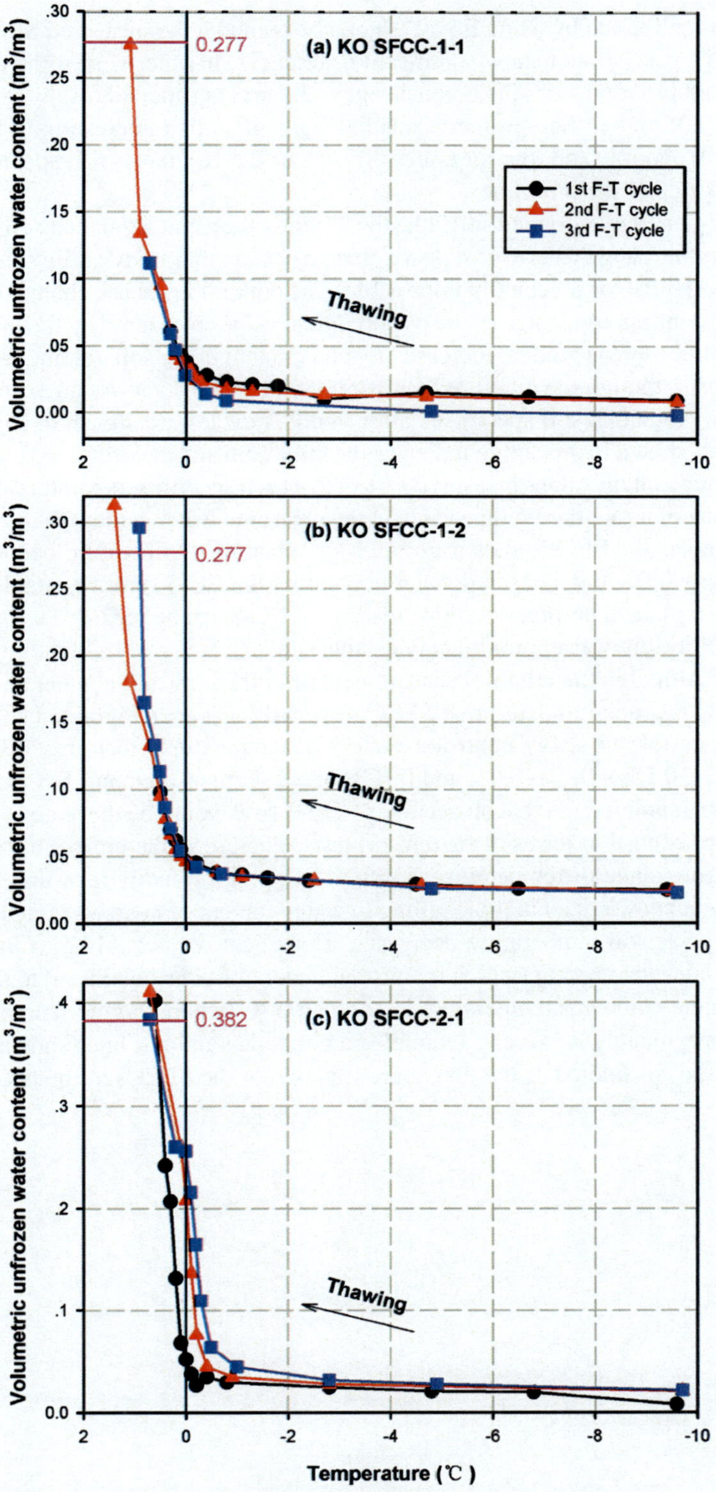

Fig. 10. Effect of F-T cycles on the SFCCs of three Komaoka specimens

was calibrated by using Eq. (1). It can be seen that the measured SFCCs under the three F-T cycles do not show significant differences. In other words, the effect of F-T cycles is not pronounced. This is consistent with the experimental results by Ren and Vanapalli (2020) on five fine-grained Canadian soils. It is also interesting to note that similar to the Toyoura sand, the measured SFCCs of the Komaoka soil specimens spread to the left side of the 0 °C line.

For coarse-grained soils like the Toyoura sand and Komaoka soil, in the narrow temperature range where most phase change occurs, thermodynamic equilibrium is difficult to establish or practically impossible. The continuous phase change in soil pores causes the continuous change of the water content values measured by the EC-5 sensor. In other words, there is gradual increase of water content in the soil volume sampled by the EC-5 sensor, though the measured temperature looks constant. As an example, the SFCCs of two Komaoka soil specimens after two F-T cycles are shown in Fig. 11. The whisker plots shown in this figure indicates the minimum and maximum (i.e. gradual increase) of water content values measured by EC-5 at each specific temperature near 0 °C. The open data points are the average value of the measured water contents at each temperature (the same as the SFCCs shown in Fig. 10). When the temperature becomes more negative below 0 °C, the measured water content by the EC-5 sensor is stable (i.e. no whisker). As explained before, the stable reading of EC-5 can be regarded as that thermodynamic equilibrium was approximately established.

Although the reliable measurement of unfrozen water content and temperature near 0 °C is crucial to determine SFCC, especially for coarse-grained soils, it is practically impossible to stably control a variety of temperatures that are close to 0 °C (e.g., −0.2, −0.15, −0.1, −0.05, and 0 °C) for SFCC measurement. For example, Kong et al. (2020) pointed out that an accuracy of ± 0.1 °C would be the state of the art for temperature control in terms of frozen soil tests. Therefore, the unfrozen water content in this narrow range of temperature is important yet not reliably measured. Since it is generally acknowledged that the unfrozen water content of coarse-grained soils dramatically drops as soil temperature decreases, a practical way could be to linearly connect the initial water content (which is constant under unfrozen condition) to the measured water content value at a temperature lower than 0 °C (where thermodynamic equilibrium can approximately achieves). Examples are the blue and red lines shown in Fig. 11, which can be considered as the first approximation of the SFCCs of the two specimens.

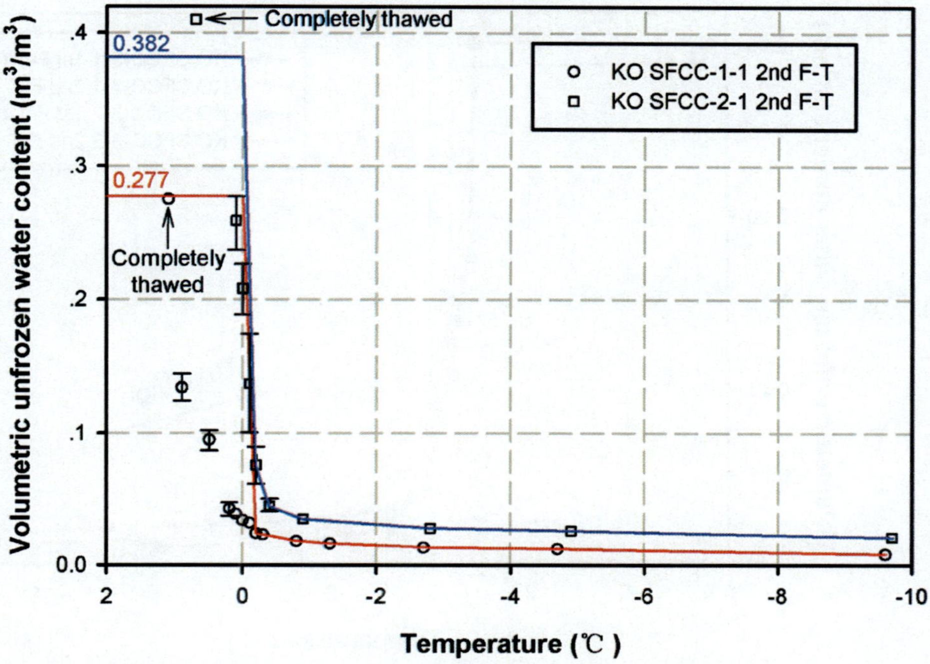

Fig. 11. The SFCCs of two Komaoka specimens after two F-T cycles

Figure 12 shows the SFCC results of two Komaoka specimens, which were respectively subjected to one- and three-dimensional F-T conditions, as shown in Fig. 4(a). The two specimens, which were initially under unfrozen condition, were directly subjected to −10 °C in each of the F-T cycles. The controlled temperature was then gradually increased to desired subzero temperatures for measuring the thawing branch of SFCC. The average values for the unfrozen water content under each of the temperatures were used to construct the SFCCs in Fig. 12. The measured water content at the completely thawed condition was calibrated by using Eq. (1). It can be seen that under one-dimensional condition, the effect of F-T cycles on SFCC is not significant, similar to the three-dimensional condition. In addition, there is not large difference between the measured SFCCs under the one- and three-dimensional conditions, although it took much longer time for the one-dimensional case to achieve steady state.

The temperature distributions in the specimen under one- and three-dimensional freezing/thawing are different. For example, in one-dimensional freezing condition, the temperature inside the specimen is not uniform. Rather, there is temperature gradient within the specimen. The temperature distribution can be considered linear.

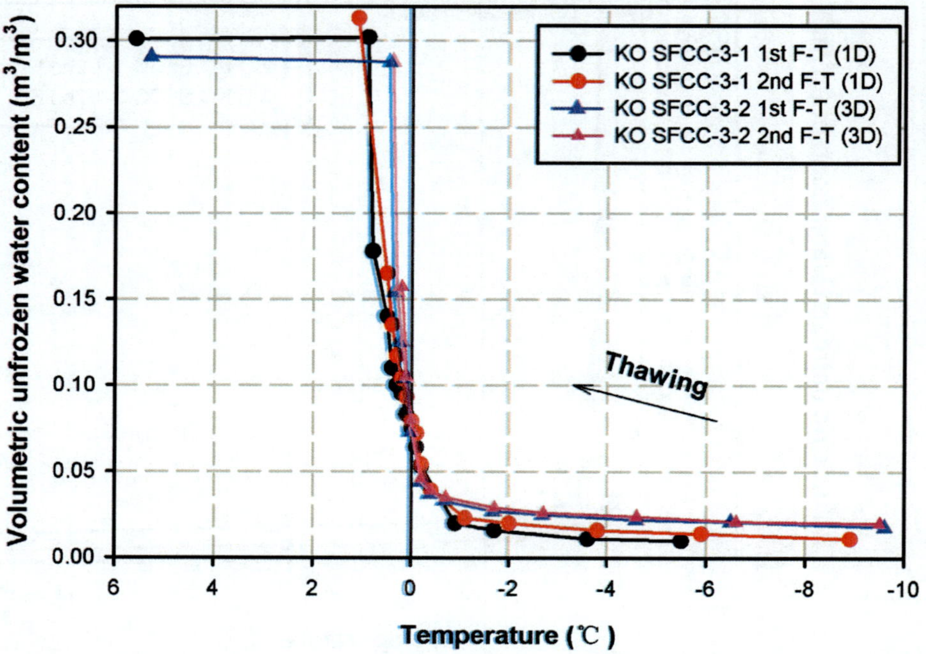

Fig. 12. The SFCCs of two Komaoka specimens subjected to one- and three-dimensional F-T conditions

4 Summary

The SFCC describes the relationship between the amount of unfrozen water and its energy state or subzero temperature in a frozen soil. It is essential to estimating the physical and mechanical properties of frozen soils and to modeling the transport mechanism of water, heat, and solutes in cold regions. In the present study, the SFCC of the Komaoka soil, which is a volcanic soil sampled in Sapporo, Hokkaido, Japan, was investigated. The SFCC test is time-consuming as relatively long time is required to approximate to thermodynamic equilibrium condition under each controlled temperature. The EC-5 moisture sensor was used to measure the volumetric water content of the specimens. Good contact between the soil body and the prongs of the EC-5 sensor is generally essential for good results. Great care should also be paid when analyzing the EC-5 measured water content.

By comparing with the SFCCs of two Toyoura sand specimens measured by NMR, it is recognized that the SFCC measured by EC-5 sensor shows good result when no ice is present or most pore water changes phase to ice in the soil specimen. However, the measurements in the temperature range where massive phase change occurs may bear errors. A couple of SFCC tests were carried out on the Komaoka soil, by considering different F-T cycles (up to three) and freezing/thawing methods (i.e. one- and three-dimensional). The measured SFCCs under three F-T cycles do not show significant differences. In other words, the effect of F-T cycles is not pronounced. Under one-dimensional condition, the effect of F-T cycles on SFCC is not significant, similar to the three-dimensional condition. In addition, there is not large difference between the measured SFCCs under the one- and three-dimensional conditions. More investigations however should be conducted to give more credits to the results of the present study.

Acknowledgement. The first author gratefully acknowledges the financial support from the Japan Society for the Promotion of Science (JSPS).

References

Anderson, D.M., Tice, A.R.: Predicting unfrozen water contents in frozen soils from surface area measurements. Highway Res. Rec. **393**, 12–18 (1972)

Azmatch, T.F., Sego, D.C., Arenson, L.U., Biggar, K.W.: Using soil freezing characteristic curve to estimate the hydraulic conductivity function of partially frozen soils. Cold Reg. Sci. Technol. **83**, 103–109 (2012)

Grant, S.A., Sletten, R.S.: Calculating capillary pressures in frozen and ice-free soils below the melting temperature. Environ. Geol. **42**(2–3), 130–136 (2002). https://doi.org/10.1007/s00254-001-0482-y

Ishikawa, T., Tokoro, T., Nakamura, D., Yamashita, S.: Influence of freeze-thaw action on air-permeability of unsaturated soil ground. In: Third International Conference on Geotechnique, Construction Materials and Environment, Nagoya, Japan, 13–15 November 2013, pp. 71–76 (2013)

Kawamura, S., Miura, S.: Rainfall-induced failures of volcanic slopes subjected to freezing and thawing. Soils Found. **53**(3), 443–461 (2013)

Kong, L., Wang, Y., Sun, W., Qi, J.: Influence of plasticity on unfrozen water content of frozen soils as determined by nuclear magnetic resonance. Cold Regions Sci. Technol. **172**, 102993 (2020)

Koopmans, R.W.R., Miller, R.D.: Soil freezing and soil water characteristic curves. Soil Sci. Soc. Am. J. **30**(6), 680–685 (1966)

Kozlowski, T.: A comprehensive method of determining the soil unfrozen water curves: 1. Application of the term of convolution. Cold Reg. Sci. Technol. **36**(1–3), 71–79 (2003)

Liu, Z., Yu, X.: Predicting the phase composition curve in frozen soils using index properties: a physico-empirical approach. Cold Reg. Sci. Technol. **108**, 10–17 (2014)

Mao, Y., Romero Morales, E.E., Gens Solé, A.: Ice formation in unsaturated frozen soils. In: Unsaturated Soils: UNSAT 2018: The 7th International Conference on Unsaturated Soils, pp. 597–602. The Hong Kong University of Science and Technology (HKUST) (2018)

Mu, Q.Y., Ng, C.W.W., Zhou, C., Zhou, G.G.D., Liao, H.J.: A new model for capturing void ratio-dependent unfrozen water characteristics curves. Comput. Geotech. **101**, 95–99 (2018)

Neall, V.E.: Volcanic soils. Land Use Land Cover Soil Sci. **7**, 23–45 (2009)

Nguyen, T.B.: Effect evaluation of grass on infiltration and seepage of volcanic soil ground. Master thesis, Hokkaido University, Sapporo, Japan (2017)

Ren, J., Vanapalli, S.K.: Comparison of soil-freezing and soil-water characteristic curves of two Canadian soils. Vadose Zone J. **18**(1), 1–14 (2019)

Ren, J., Vanapalli, S.K.: Effect of freeze–thaw cycling on the soil-freezing characteristic curve of five Canadian soils. Vadose Zone J. **19**(1), e20039 (2020)

Schafer, H.: Freezing characteristics of mine waste tailings and their relation to unsaturated soil properties. Master thesis, University of Alberta, Edmonton, Canada (2018)

Smith, M.: Models of soil freezing. In: Church, M., Slaymaker, O. (eds.) Field and Theory: Lectures in Geocryology, pp. 96–120. University of British Columbia Press, Vancouver (1985)

Spaans, E.J., Baker, J.M.: The soil freezing characteristic: its measurement and similarity to the soil moisture characteristic. Soil Sci. Soc. Am. J. **60**, 13–19 (1996)

Watanabe, K., Osada, Y.: Comparison of hydraulic conductivity in frozen saturated and unfrozen unsaturated soils. Vadose Zone J. **15**(5), 1–7 (2016)

Watanabe, K., Wake, T.: Measurement of unfrozen water content and relative permittivity of frozen unsaturated soil using NMR and TDR. Cold Reg. Sci. Technol. **59**(1), 34–41 (2009)

Xu, J., Lan, W., Li, Y., Wang, S., Cheng, W.C., Yao, X.: Heat, water and solute transfer in saline loess under uniaxial freezing condition. Comput. Geotech. **118**, 103319 (2020)

Yu, L., Zeng, Y., Wen, J., Su, Z.: Liquid-Vapor-Air Flow in the frozen soil. J. Geophys. Res. Atmos. **123**(14), 7393–7415 (2018)

Zhang, S., et al.: Canopy effect caused by vapour transfer in covered freezing soils. Géotechnique **66**(11), 927–940 (2016)

Structural Characterization of Residual Soil and the Effect of Drying and Wetting Cycles on Its Strength

Wei Bai[1(✉)], Lingwei Kong[1], Liming Xu[2], Xianwei Zhang[1], Zhiliang Sun[1], Ran An[1,3], and Xiu Yue[1,3]

[1] State Key Laboratory of Geomechanics and Geotechnical Engineering, Institute of Rock and Soil Mechanics, Chinese Academy of Sciences, Wuhan, Hubei, China
william_bai@yeah.net
[2] Xiamen Rail Transit Group Limited Corporation, Xiamen, Fujian, China
[3] University of Chinese Academy of Sciences, Beijing, China

Abstract. A series of indoor tests was performed to obtain the physical and mechanical properties and structural characterization of granite residual soil. Experimental results indicated that clot and flocculation are the primary forms of microstructure, and the residual appearance of the parent rock's crystal pattern is reserved. Structural strength is provided by strong cementation and residual chemical bond force. The clay mineral is mostly kaolin, which exhibits a laminated structure. The grain particle of granite residual soil presents the mixed features of sand and clay. The coefficient of compressibility is mostly distributed with in the range of 0.30–0.35 MPa^{-1}, which is for medium compressible soil. Granite residual soil demonstrates good mechanical performance even under saturation condition. However, strength variability is notable. Drying and wetting cycles reduce the strength of granite residual soil, and strength decreases by approximately 30% after two drying and wetting cycles.

Keywords: Residual soil · Physical properties · Strength · Drying and wetting cycles

1 Introduction

Granite residual soil is formed by the physical and chemical weathering of a rock in a hot and humid environment; such soil overlays its parent rock (Shang et al. 2014, 2013). Granite residual soil exhibits fortissimo structural properties. It generally contains a large amount of coarse sand and gravel. In addition, it demonstrates the characteristics of high void ratio, high liquid limit, high strength, and low to medium compressibility. Given the lack of understanding of the characteristics of granite residual soil, several engineering accidents, such as subgrade slope instability and pile foundation failure (Cheng 2002; Liu 1999; Zou and Li 2002), can occur. The distribution area of granite eluvial deposits accounts for approximately 2/3 of the land area of Xiamen (Wang et al. 1990; and Chen and Gong 2014), China. Granite weathered strata, with a thickness of 10–50 m, are

Y. Liu et al. (Eds.): GeoChina 2021, SUCI, pp. 97–111, 2021.
https://doi.org/10.1007/978-3-030-80316-2_10

widely distributed along the metro lines of Xiamen. They compose the major rock and soil layer involved in the construction of foundation pits and other engineering structures. The evident seasonal drying and wetting cycles and the frequent fluctuations of ground water level exert considerable impact on the strength of the granite residual soil in this area. Hence, the acquisition and evaluation of the engineering geological characteristics, material composition, microstructure, and physical and mechanical properties of granite residual soil are highly significant in engineering to avoid accidents and fully utilize the bearing potential of this type of soil.

Therefore, the mineral composition, microstructure characteristics, diameter, grading distribution, drying and wetting cycle effect, and strength parameters of Xiamen granite residual soil were tried and tested in the current study. The engineering geological characteristics of this soil type were also analyzed and evaluated accordingly to provide references for engineering constructions and geological explorations in this area.

2 Engineering Geological Characteristics

Samples were collected from three sites along the Xiamen metro lines located near Xingjin Station (XJ), Lvcuo Station (LC), and Yuanboyuan Station (YB). Through the combination of drilling, standard penetration test (SPT), and field investigation, the soil layer distribution in XJ was determined as follows: ① 0–3.2 m, mixture of clay and gravel; ② 3.2–7.2 m, plastic residual sandy clay with mixed colors of gray–white and yellow–brown, quartz particles smaller than 3 mm in diameter; ③ 7.2–23.0 m, hard plastic residual sandy clay with mixed colors of gray–white and yellow–brown, the content and size of quartz particles increase with depth; ④ 23–27.2 m, completely weathered granite residual soil with yellow–brown color and hard texture, quartz particles are cemented; ⑤ 27.2–30.5 m, moderately to strongly weathered granite residual soil, quartz particle content is smaller compared with that in the upper layer. The soil layer distributions of LC and YB are similar to that of XJ. The relationship between depth and SPT in the three sites is illustrated in Fig. 1. The standard penetration number, $N_{63.5}$, gradually increases with an increase in depth. The physical state changes from plastic to slightly plastic to nonplastic. The strength and foundation bearing capacity gradually increase. A typical distribution pattern of granite strata is presented: the weathering and clayization degree of granite gradually weaken from the shallow portion to the deep region.

Open-pit exploration was performed, and cubic soil samples with dimensions of 25 cm × 25 cm × 25 cm were manually collected from 6 m to 7 m below the surface of the three sites, as shown in Fig. 2(a). No groundwater was found during the process. The samples were packed to make them shock resistant and easy to transport, as shown in Fig. 2(b).

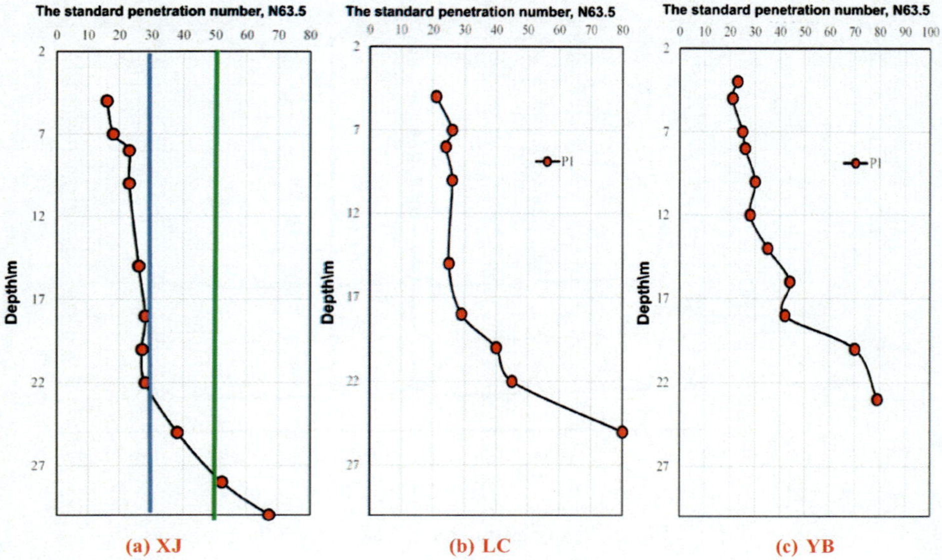

Fig. 1. Relationship between depth and SPT

Fig. 2. Process of taking the undisturbed sample

3 Composition

Undisturbed soil was ground and passed through a 200-mesh sieve. Mineral composition was analyzed via X-ray diffraction, as shown in Fig. 3. The proportions of kaolin, illite, and quartz in XJ are 37.42%, 9.2%, and 53.38% respectively. The proportions of kaolin and quartz in LC are 24.69% and 75.31%, respectively. Kaolin and quartz in YB account for 33.03% and 66.97%, respectively. The proportion of quartz in Xiamen granite residual soil is the highest, and kaolin is the major clay mineral.

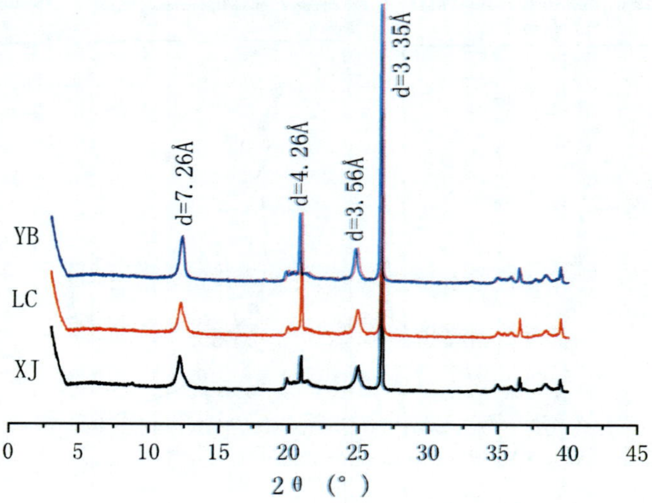

Fig. 3. X-ray diffraction patterns of granite residual soils

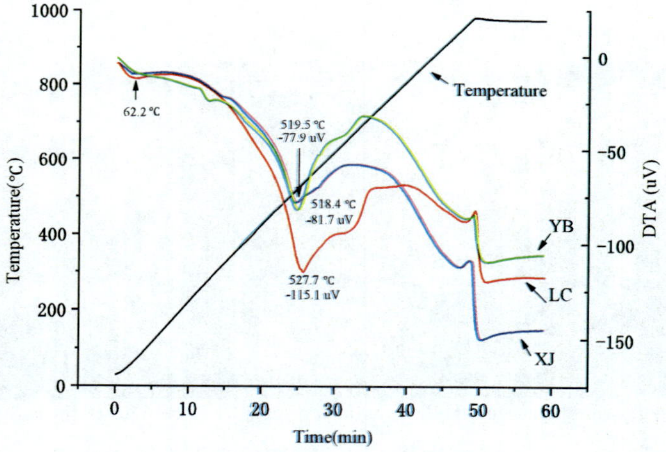

Fig. 4. DTA curves of granite residual soils

The heat absorption valley and exothermic peak in the differential thermal analysis (DTA) curves are caused by the thermal effects of dehydration and phase change during heating. They are related to the composition and structure of granite residual soil. The DTA curves of granite residual soil in XJ, LC, and YB are shown in Fig. 4.

During the temperature increase period of the aforementioned granite residual soil, a small endothermic reaction valley appeared in the DTA curves when temperature reached approximately 62.2 °C. This valley represents the loss of adsorbed water on the particle surface. When the temperature reached approximately 520 °C, the DTA curves presented a large endothermic reaction valley, indicating that a huge amount of adsorbed water escaped within this temperature range and the crystal structure was destroyed during this period. To a certain extent, this endothermic reaction valley can be used to reflect the water absorption capacity of the soil. In general, granite residual soil with a large endothermic reaction valley has high moisture content and viscosity (Zhang 1989). By comparing the DTA curves of the three sites, the endothermic reaction valley of the LC soil sample was found to be the widest. Therefore, the clay particle content of this soil sample is relatively high. The preceding conclusions can be verified in the subsequent study on diameter and grading distribution.

4 Structural Characterization

As shown in Fig. 5, quartz particles are wrapped with clay in the soil samples collected from the three sites. The scanning electron microscopy (SEM) image magnified 100× clearly shows that the quartz particles are lumpy and surrounded by oxides or other cementitious materials in clay Fig. 6(a).

(a) XJ (b) LC (c) YB

Fig. 5. Digital photographs of granite residual soils

When zoomed in to 800× times, the contact mode between particles is found to be dominated by edge-to-face or face-to-face mosaic contact, and the cracks are clear Fig. 6(b). The structure of clay minerals between quartz particles is nondirectional stacking, as shown in Fig. 6(c). This type of structure is attributed to the parent rock gradually weathering into kaolin in an acidic environment. It retains the internal structure characteristics of the primary minerals and the residual chemical bond strength between the grains of the primary minerals. The latter is one of the important reasons for the high mechanical strength of this type of structure. When the clay minerals are enlarged 2000×, the accumulation place of kaolin mineral particles was found to be flat, as shown in Fig. 6(d). This is a typical laminated domain structure of residual soil dominated by kaolin minerals.

From the discussion in the preceding paragraph, the quartz particles in granite residual soil are wrapped by or filled with cement; thus, cementation occurs between particles. However, tests and theoretical studies on soil structure have focused on arrangement characteristics for a long time, but have disregarded the importance of intergranular connection characteristics (Zhang 1994; Wang et al. 2000; Shang et al. 2004). These cementing materials are generally free oxides (Shang et al. 2013) that are sensitive to

(a) quartz particles at 100×

(b) particle contact mode at 800×

(c) stacking structure of minerals at 800×

(d) clay minerals at 2000×

Fig. 6. The SEM photographs of granite residual soils

water content. Mechanical strength easily and abruptly changes under moisture content increase or drying and wetting cycle. Such change is one of the important reasons for the evident structural sensitivity of granite residual soil.

5 Physical Properties

The physical indexes of Xiamen granite residual soil are listed in Table 1. Granite residual soil in the three sites generally have higher natural moisture content and larger void ratio. However, their natural state is basically nonplastic or slightly plastic. The liquid limits of the fine particle group of the three aforementioned granite residual soil samples are all higher than 50%, and their plasticity indexes are all greater than 22. In the plastic index diagram, XJ and YB are located slightly lower than Line A (MH) but LC is located slightly higher than Line A (CH). This finding indicates that the fine particle group of Xiamen granite residual soil exhibits the properties of high liquid limit clay and silty clay.

The particle composition of the granite residual soil mentioned above is special. The quality of particles with a diameter greater than 0.075 mm is not more than 50% of the total quality, and the plasticity index is greater than 17. Thus, such type of soil should be classified as clay on the basis of standards (Ministry of Construction 2009). However,

Table 1. Indexes of physical properties of granite residual soils

Soil sample	XJ	LC	YB
Depth (m)	6.0	6.5	7.0
Average moisture content (%)	23.4	31.0	25.7
Wet density (g/cm^3)	1.84	1.76	1.82
Dry density (g/cm^3)	1.49	1.34	1.45
Liquid limit (%)	58.5	51.8	51.4
Plastic limit (%)	31.4	27.9	28.6
Plasticity index	27.1	23.9	22.8
Specific gravity	2.7	2.7	2.7
Void ratio	0.81	1.01	0.86
Saturation degree (%)	77.9	82.9	80.2
Particle composition (%)			
10~5 mm	14.3	0.4	4.2
5~2 mm	13.8	14.3	20.4
2~0.5 mm	10.7	12.7	12.9
0.5~0.25 mm	4.4	6.2	5.2
0.25~0.075 mm	3.6	7.0	3.0
0.075~0.005 mm	39.3	42.8	29.8
<0.005 mm	13.9	16.6	24.5

the quality of particles with a diameter greater than 2 mm is close to or more than 25% of the total quality. Clay does not reflect such characteristics, and thus, classifying such soil as *residual gravel clay* is more appropriate in reference with the local standards (Fujian Provincial Department of Construction 2006). The proportion of coarse particles and fine particles is large, while the content of particles with intermediate particle size is small. Moreover, the aforementioned characteristics of particle diameter distribution provide possible conditions for small particles to flow out from the pores of large particles. When hydrodynamic pressure is too high, seepage deformation phenomena, such as piping and flowing soil, may easily occur. Therefore, effective water-sealing measures should be implemented in important projects. The data in Table 1 indicate that the proportion of the fine particle group of the LC soil samples reaches as high as 59.4%, exceeding those of the XJ (53.2%) and YB (54.3%) soil samples. This result is consistent with the conclusion that the LC soil samples with the widest endothermic reaction valley have a higher content of viscous particles in the DTA curves. Permeation tests were also conducted on 10 samples from each site. The change range of the XJ soil samples' permeability coefficient is 1.01×10^{-5}–3.90×10^{-4} cm·s^{-1}. The change range of the LC soil samples' permeability coefficient is 7.43×10^{-6}–9.07×10^{-5} cm·s^{-1}. The change range of the YB soil samples' permeability coefficient is 8.11×10^{-6}–$1.66 \times$

10^{-4} cm·s^{-1}. The histogram of the permeability coefficient of the three sites is presented in Fig. 7. The probability that the permeability coefficient of the sample is 10^{-5} cm·s^{-1} is the highest. Therefore, they belong to a weak permeable grade.

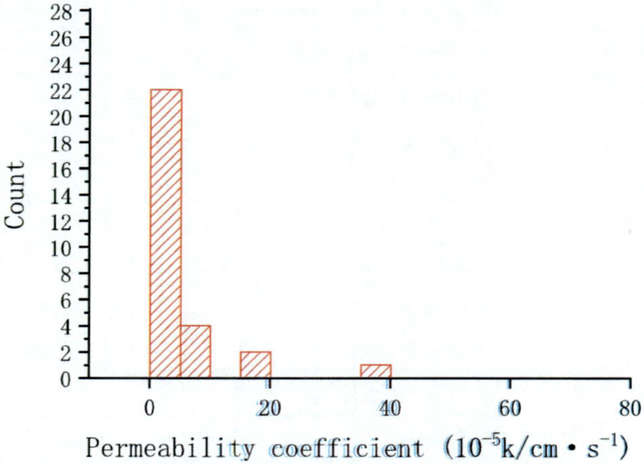

Fig. 7. Histogram of permeability coefficient

6 Mechanical Properties

6.1 Compressibility

Undisturbed soil samples were cut useing a ring knife (61.8 mm in diameter and 20 mm in height). 10 samples were collected from each site for the standard consolidation tests. The histogram of the compressibility coefficient is presented in Fig. 8.

As shown in the figure, the compressibility coefficient (a_{1-2}) of each sample is between 0.15 MPa^{-1} and 0.45 MPa^{-1}. Most of the values are within the range of 0.30–0.35 MPa^{-1} and can be classified as medium compressibility soil. The compressibility of XJ and YB is lower than that of LC.

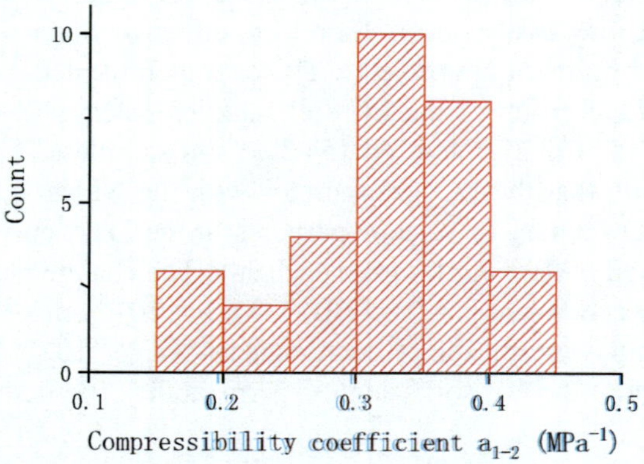

Fig. 8. Histogram of compressibility coefficient

6.2 Consolidated Undrained (CU) Triaxial Tests

Undisturbed soil samples were cut into cylindrical soil samples (diameter: 39.1 mm, height: 80 mm) in accordance with the standard (Ministry ofWater Resources 2019). Triaxial compression tests (CU and consolidated shear tests) were performed after reaching saturation (vacuum pumping saturation method). The shear rate was set as 0.060 mm/min. Nine to ten tests were conducted at each site. The typical failure modes of the soil samples from the three sites after the CU tests are shown in Fig. 9. No obvious shear plane appeared when axial strain reached 20%.

Fig. 9. The failure mode of granite residual soil samples after the CU tests

The CU shear strength indexes (cohesion and internal friction angle) obtained in the preceding tests are provided in Table 2.

Table 2. Consolidated undrained triaxial tests results of granite residual soils

Soil sample	c_{cu}/kPa	φ_{cu}/°	Soil sample	c_{cu}/kPa	φ_{cu}/°	Soil sample	c_{cu}/kPa	φ_{cu}/°
XJ1	33.1	25.1	LC1	64.2	15.3	YB1	22.3	29.9
XJ2	56.5	22.7	LC2	40.7	19.8	YB2	74.7	19.3
XJ3	44.0	23.5	LC3	61.4	13.8	YB3	83.3	15.9
XJ4	54.1	19.8	LC4	–	–	YB4	43.8	19.0
XJ5	–	–	LC5	38.5	18.8	YB5	49.4	22.0
XJ6	41.9	23.1	LC6	60.0	14.0	YB6	24.5	23.6
XJ7	59.5	22.3	LC7	46.6	18.1	YB7	24.5	23.6
XJ8	61.2	22.5	LC8	84.8	13.1	YB8	21.3	30.1
XJ9	55.6	22.6	LC9	58.7	17.3	YB9	74.2	20.1
XJ10	51.7	23.8	LC10	59.2	16.2	YB10	78.4	19.6

The c_{cu} of XJ ranges from 33.1 kPa to 61.2 kPa. Its mean value is 50.8 kPa, its standard deviation is 8.8 kPa, and its variable coefficient is 0.17. The φ_{cu} of XJ ranges from 19.8° to 25.1°. Its mean value is 22.8°, its standard deviation is 1.3°, and its variable coefficient is 0.03. The c_{cu} of LC ranges from 38.5 kPa to 84.8 kPa. Its mean value is 57.1 kPa, its standard deviation is 13.2 kPa, and its variable coefficient is 0.23. The φ_{cu} of LC ranges from 13.1° to 19.8°. Its mean value is 16.3°, its standard deviation is 2.2°, and its variable coefficient is 0.14. The c_{cu} of YB ranges from 21.3 kPa to 83.3 kPa. Its mean value is 49.6 kPa, its standard deviation is 24.6 kPa, and its variable coefficient is 0.50. The φ_{cu} of YB ranges from 13.1° to 19.8°. Its mean value is 16.3°, its standard deviation is 2.2°, and its variable coefficient is 0.14. The data of the strength parameters of the CU and consolidated shear tests indicate that even in a saturated state, the granite residual soil samples exhibit higher cohesion and internal friction angle and still have better mechanical properties. However, although the three sites are basically located in the same soil layer, the quality of the soil samples is not uniform, presenting a highly variable coefficient of cohesion. Xiamen residual soil is the debris formed by the weathering of a rock. Thus, variation in mechanical parameters will be caused by different forms of primary and secondary structural surfaces, particle size distribution among soil samples, and connection patterns of wrapping and filling oxide. Therefore, the inhomogeneity of residual soil should be considered fully in practical projects, and parameters should be selected on the basis of in-situ test data as much as possible in major projects.

6.3 Drying and Wetting Cycles

Undisturbed soil samples were cut into cylindrical soil samples (diameter: 39.1 mm, height: 80 mm) in accordance with the standards. The drying and wetting cycle range was between 15% and 30% (saturated moisture content was approximately 30%; it slightly varied for soil samples from different sites). Humidification was performed using the standard method (i.e., vacuum pumping saturation). Dehumidification was conducted via air-drying: the samples were removed from the saturator, placed on a tray, and air-dried to the specified moisture content (monitored by weighing). To ensure the uniformity of moisture concent in the soil sample after air drying, the soil sample must be sealed with plastic film and then placed in a moisturizing tank for 24 h. The specific process of the drying and wetting cycle tests is illustrated in Fig. 10. Four sets of samples were prepared for each site, for a total of twelve sets of samples. The unconfined compressive strength indexes were obtained using a YYW-2 unconfined compression tester.

The test results of the samples from the three sites were analyzed. The decay rate is defined as the ratio of the difference of the unconfined compressive strength between samples of n cycles and samples of 0 cycles to the strength corresponding to 0 cycles, as shown in Eq. (1).

$$\Delta = \left| \frac{q_{un} - q_{u0}}{q_{u0}} \right| \times 100\% \tag{1}$$

Where, q_{un} is the unconfined compressive strength of the soil samples that have undergone n-th drying and wetting cycles, and q_{u0} is the unconfined compressive strength of the soil samples that have not undergone a drying and wetting cycle.

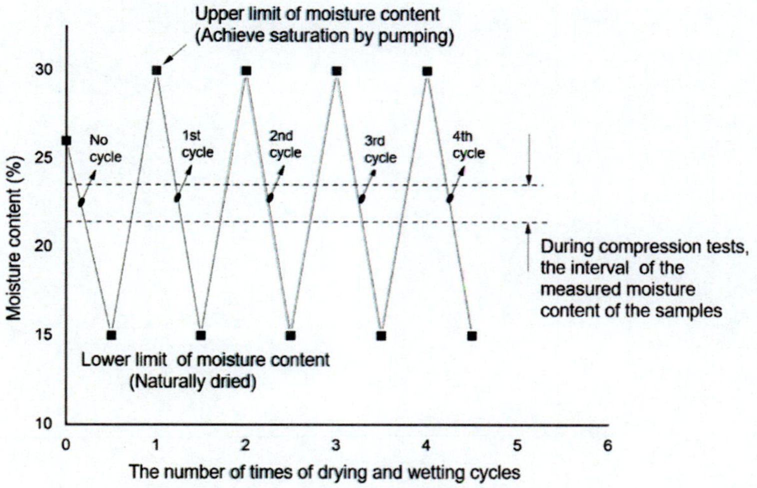

Fig. 10. The sketch of the drying and wetting cycle

The relationship among unconfined compressive strength, decay, and drying and wetting cycles is shown in Fig. 11. The statistical test results are provided in Table 3.

As indicated in Fig. 11 and Table 3, unconfined compressive strength decreases rapidly after a drying and wetting cycle. The attenuation amplitude of the XJ soil sample is the largest (30.6%), followed by that of the YB soil sample (22.1%) and then the LC soil sample (19.1%). After a second drying and wetting cycle, the attenuation amplitude of the XJ and YB soil samples is approximately 10%, except for the large reduction (46.2%) in LC. The change amplitude of all the soil samples in the third and fourth cycles is less than 4%. Therefore, a conclusion can be drawn that the drying and wetting cycle considerably influences the unconfined compressive strength of granite weathered soil samples. In addition, the unconfined compressive strength tends to become stable after two drying and wetting cycles.

Xiamen granite residual soil is mostly composed of residual clastic materials formed in situ after the weathering of a rock mass. Most of the rock is completely weathered into mineral particles. The common feature of residual soil and weathered rocks, except for isolated rocks, is that they remained in the location of the original rock without being transported and sorted by other media. There must be primary and secondary structural planes in the residual soils, and the structure of the original rock is relatively preserved between the mineral particles. The drying and wetting cycle affects the structure of residual soil, generating more microcracks and reducing the strength of soil. Moreover, the properties of cementing materials irreversibly weaken during the drying and wetting cycle, reducing the structural strength of granite residual soil.

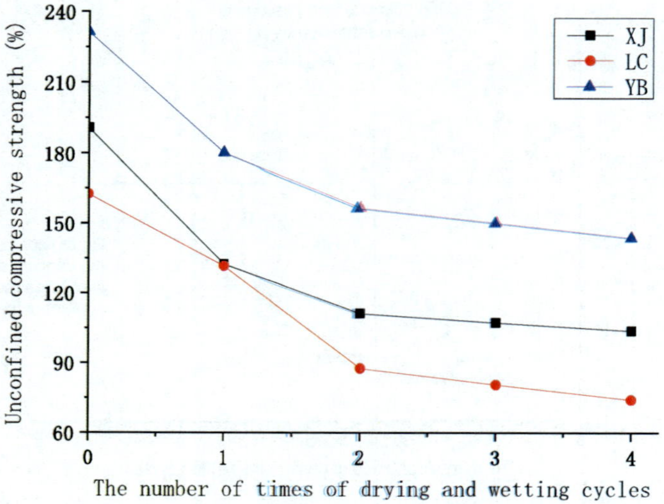

(a) Relationship between unconfined compressive strength and drying and wetting cycles

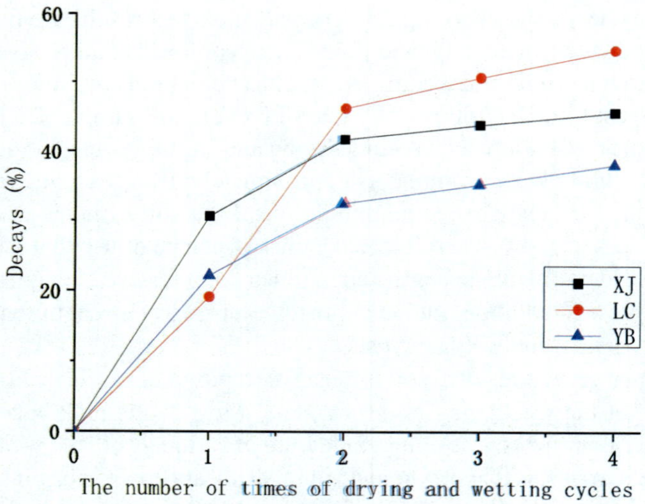

(b) Relationship between decays and drying and wetting cycles

Fig. 11. Relationship between unconfined compressive strength, decays and drying and wetting cycles

Table 3. The unconfined compression tests results after drying and wetting cycles

Soil samples	Number of times of cycles	Unconfined compressive strength (kPa)	Moisture content (%)	Decays (%)
XJ	0	190.8	20.7	0.0
	1	132.4	20.1	30.6
	2	111.2	20.0	41.7
	3	107.2	20.2	43.8
	4	103.7	20.5	45.6
LC	0	162.6	22.1	0.0
	1	131.6	22.0	19.1
	2	87.5	22.5	46.2
	3	80.4	22.3	50.6
	4	73.9	22.6	54.5
YB	0	231.5	20.8	0.0
	1	180.3	20.5	22.1
	2	156.4	20.9	32.5
	3	150.0	20.1	35.2
	4	143.6	22.6	38.0

7 Conclusions

A series of tests were applied to determine the physical, mechanical and structural characteristics of the typical granite residual soil in Xiamen, China, namely, drilling, standard penetration tests (SPT), open-pit exploration, X-ray diffraction, differential thermal analyses (DTA), scanning electron microscopy (SEM), liquid limit and plastic limit tests, particle diameter analyses, permeability tests, standard consolidation tests, consolidated undrained (CU) triaxial tests and drying-wetting cycle tests. On the basis of the obtained results, the following conclusions can be drawn:

(1) The special properties of Xiamen granite residual soil are closely related to its composition and structure. Its structural characterization retains the residual form of the crystal structure of its parent rock, primarily in the form of clot and flocculation. Structural strength mostly originates from the strong cementing force of oxide and the remaining chemical bonding force. Quartz particles are wrapped by or filled with cement. The edge-to-surface or surface-to-surface inlay contact is the major contact mode between particles. The composition of clay minerals is largely kaolin, which exhibits a typical laminated domain structure.

(2) The particle composition of Xiamen granite eluvial soil is special. It presents the characteristics of sand, gravel, and clay. The characteristics of particle diameter distribution provide possible conditions for small particles to flow out from the

pores of large particles. For important projects, such as foundation pits, effective water-sealing measures should be implemented to avoid seepage deformation, such as piping and flowing soil.

(3) The coefficient of compressibility of granite residual soil from the three sites in Xiamen is mostly distributed within the range of 0.30–0.35 MPa^{-1}, which is for medium compressible soil. The compressibility of XJ and YB is lower than that of LC. The CU shear tests show that Xiamen granite residual soil exhibits better mechanical performance even under saturated condition. However, strength variability is notable, and nonuniformity should be considered fully in practical engineering.

(4) The drying and wetting cycle affects the structure of granite residual soil. It generates more microcracks and irreversibly weakens the properties of the cementing materials and structural connections, leading to the strength attenuation of the granite residual soil. After two drying and wetting cycles, the decay rate of unconfined compressive strength is over 30%. Thereafter, the influence of increasing the number of drying and wetting cycles on unconfined compressive strength is decreased.

(5) The above conclusions can provide valuable references for engineering construction and engineering geological exploration, and are of great engineering significance for avoiding accidents and making full use of the bearing potential of granite residual soil.

Acknowledgments. The project are funded by the National Natural Science Foundation of China (Grant No. 41772339).

References

Cheng, J.: The failure reasons of manually excavated pile in the residual area. Geotech. Eng. World (02), 36–37 (2002)

Chen, D., Gong, X.: Experiment and modeling of soil-water characteristic curve of unsaturated residual soil. Rock Soil Mech. **35**(07), 1885–1891 (2014)

Fujian Provincial Department of Construction: Local standards for engineering construction in Fujian Province: code for investigation of geotechnical engineering (2006)

Liu, D.: The failure case study of hand-dug pile foundation in Xiamen. Geotech. Eng. Tech. (04), 59–62 (1999)

Ministry of Construction, PRC: Code for Investigation of Geotechnical Engineering. China Construction Industry Press, Beijing (2009)

Ministry of Water Resources, PRC: Standard for geotechnical testing method, China (2019)

Shang, L., Sang, H., Hou, T., et al.: Experimental study on tensile strength of granite residual soil. ACTA Scientiarum Naturalium Univ. Sunyatseni **53**(06), 98–105 (2014)

Shang, L., Sang, H., Song, J., et al.: Research on soil particle joint function and brittle-elastoplastic cement damage model of unsaturated granite residual soil. Rock Soil Mech. **34**(10), 2877–2888 (2013)

Shang, Y., Wang, S., Yue, Z., et al.: Variation features of pore radius and particle diameter distributions and mineral content of completely decomposed granite and correlation of parameters. Rock Soil Mech. **25**(10), 1545–1550 (2004)

Wang, Q., Jiang, H., Tang, D.: A study on engineering geological properties of granitic residual soil in south Fujian. Geol. Fujian **9**(02), 90–99 (1990)
Wang, Y., Xie, X., Wang, C.: Physical properties of the profiles of weathered granitic rocks in Guangdong. Trop. Geogr. (04), 256–260 (2000)
Zou, J., Li, Z.: Research on the stability of completely decomposed granite cut slope. Hunan Commun. Sci. Technol. (01), 25–26+38 (2002)
Zhang, Q.: A study on the relationship between DTA curves and physico-chemical properties of bentonite. ACTA Mineral. Sin. (02), 177–180 (1989)
Zhang, W.: Analysis of the shear strenth of granitic residual soil and the destablization of soil slope. Hydrogeol. Eng. Geol. (03), 41–43 (1994)

Analysis of Tunnel Face Stability of Shield Tunneling Through Water-Rich Sand-Pebble Stratum

Wei Wang, Fan Liu(✉), Zhao Han, Xinyuan Zhang, and Xu Zhou

School of Civil Engineering, Central South University, Changsha 410075, China

Abstract. The sand and pebble stratum generally exists in the urban underground engineering construction. When disturbed by the outside world, especially under the condition of rich water, it is easy to cause stratum loss due to the characteristics of sand pebble stratum, resulting in instability of the face and surface settlement and other problems. In this paper, the physical and mechanical parameters of sand pebble soil in a subway area of Changsha were obtained through laboratory tests, and then the micro parameters were calibrated. According to the calibrated micro parameters, a PFC^{3D} full scale model considering seepage was established for the stability analysis of the tunnel face. Through the analysis of the numerical simulation results, the influence of different factors such as stress ratio, buried depth (C), shield diameter (D), and boulders on the stability of the tunnel face was obtained, and the variation characteristics of the limit support pressure and the critical stress ratio of the instability index were obtained under different conditions. The results show that the formation stability increases with the increase of stress ratio in the process of shield tunneling. When the stress ratio is less than the critical stress ratio, the tunnel face will be unstable. When the boulder is broken, the instantaneous velocity of the particle in the center of the tunnel surface can reach 3m/s, and the risk of instability of the tunnel surface is increased. The pressure of the limit support increases with the increase of the buried depth. When $C/D = 3$, the pressure of the limit support tends to be stable. The critical stress ratio decreases with the increase of C/D, and the lower the critical stress ratio is, the stronger the self-stabilizing ability of the tunnel surface is.

Keywords: Shield tunnel · Sandy pebble stratum · Micro analysis · Face stability

1 Introduction

As the main construction method of underground traffic structure, shield is widely used in urban rail transit engineering. In recent years, during the construction of underground projects in many cities, sand and pebble stratum have often been encountered. Due to the large gaps between the particles, small cohesion, large permeability coefficient and high sensitivity, when subjected to external disturbances, it is easy to form a point-to-point force transmission state between the particles. Therefore, for tunnel construction under this stratum, the stability of the face is difficult to control, and related engineering accidents are common.

Y. Liu et al. (Eds.): GeoChina 2021, SUCI, pp. 112–131, 2021.
https://doi.org/10.1007/978-3-030-80316-2_11

Regarding the tunnel face stability of shield tunneling through sand-pebble stratum, scholars have carried out researches on theoretical analysis, laboratory tests and numerical simulations. In terms of numerical simulation, finite element and discrete element are mainly two analysis methods. In terms of finite element analysis, Vermeer et al. [1] studied the relationship between the ultimate supporting force at the tunnel face and the soil parameters (internal friction angle, etc.) under drainage conditions; Zhang et al. [2] proposed a new shield tunnel instability Mechanism, and analyzed the influence of tunnel depth and soil properties on the stability of the tunnel face. In terms of discrete element analysis, Melis et al. [3–6] with the help of PFC^{3D}, studied the change law of the excavation parameters of the earth pressure shield and the stability of the tunnel face, and further studied the stability of the tunnel face during the sand excavation process of the earth pressure shield tunnel; KIRSCHA et al. [7–9] studied the related factors of tunnel face instability and their influence sensitivity; Some scholars used discrete element software to establish the sand-pebble mesoscale model, studied the stability of the tunnel face in the process of shield tunneling, and compared the calculation results with theoretical analytical values [10–13]. However, in terms of the instability of palm surface, the seepage action of water is not considered in most relevant studies, and the sand pebble formation soil particles are greatly simplified, which makes the sand pebble formation mesoscale model may be greatly different from the actual situation.

Based on the Changsha Metro Line 3, according to the parameters of sandy pebble soil obtained in laboratory tests, aiming at the deformation characteristics of sandy pebble soil in mesoscale during shield tunneling, this paper establish a multi-phase mesoscale simulation model of sandy pebble soil considering gradation characteristics and permeability to analyze the change characteristics of sandy pebble soil in front of palm during shield tunneling. This study did not aim to supply quantitative application for specific construction sites, but paid more attention to a general investigation of the behavior about tunnel face stability of shield tunneling through water-rich sand-pebble stratum.

2 Laboratory Test Analysis

2.1 Particle State Analysis of Sandy Pebble Soil

Five sandy pebble samples were selected from different locations within the scope of the project, and the statistical samples are shown in Fig. 1. Each sample contains 30 randomly selected particles. Measure the long diameter (a), medium diameter (b) and short diameter(c) of each sandy pebble. According to the measurement results, calculate the flatness (F) and sphericity (Ψ) of each group of sand and gravel particles in turn. The data results are as follows Table 1.

$$F = \frac{\mathrm{a} + \mathrm{b}}{2\mathrm{c}} \tag{1}$$

$$\psi = \frac{\sqrt[3]{\mathrm{abc}}}{\mathrm{a}} \tag{2}$$

Fig. 1. Statistical pattern of sand pebble size

Table 1. Particle state characteristics of each sample

Particle state	Sample number				
	1#	2#	3#	4#	5#
F	1.54	1.57	1.44	1.55	1.53
Ψ	0.76	0.75	0.79	0.75	0.76
F/Ψ	2.03	2.09	1.82	2.07	2.01

It can be seen from Table 1 that the flatness and sphericity are between 1.44–1.57 and 0.75–0.79 respectively, and the ratio of them is basically stable at about 2, indicating that the morphological differences of each sample are very small and most of them are oblate bodies.

2.2 Sandy Pebble Soil Screening Experiment

The grading curve of sandy pebble soil samples taken in this research is shown in Fig. 2. It can be seen from Fig. 2 that the particle size can be roughly divided into three groups: 0–20 mm, 20–40 mm, and 40–80 mm, each accounting for about 1/3.

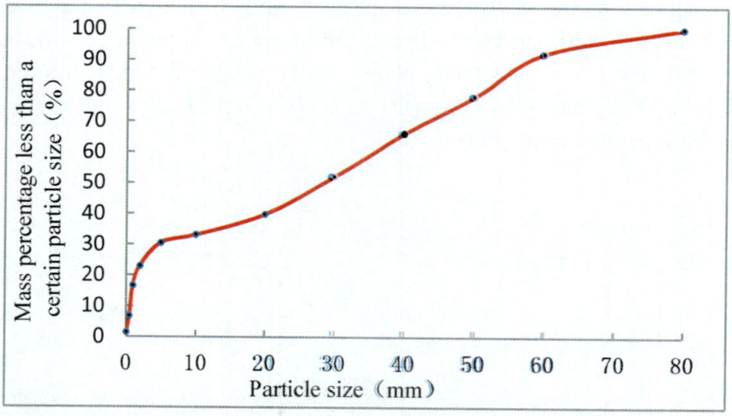

Fig. 2. Grading curve of sandy pebble soil sample

2.3 Sandy Pebble Soil Permeability Test

In this paper, 15 groups of sandy pebble soil with different gradations are carried out and analyzed to study the relationship between effective size (d_{10}), uniformity coefficient (C_u), curvature coefficient (C_c) and permeability coefficient, and then obtain the empirical formula of permeability coefficient and the three.

(1) Effect of d_{10} on permeability coefficient (K_1)

If Cu and Cc are constant, d10 becomes the only variable that affects the permeability coefficient. So in this part we take Cu = 28.5 and Cc = 0.64, the specific working condition design is shown in Table 2 below.

Table 2. Interval content of each particle size group under different d_{10}

Particle size (mm)	$C_u = 28.5$ $C_c = 0.64$				
	$d_{10} = 0.175$ mm	$d_{10} = 0.275$ mm	$d_{10} = 0.375$ mm	$d_{10} = 0.5625$ mm	$d_{10} = 0.75$ mm
0.1–0.25	10	5	5	3	2
0.25–0.5	20	15	10	5	5
0.5–1.0	30	25	20	15	10
1.0–2.0	40	40	29	20	20
2.0–5.0	50	50	40	31	29
5.0–10.0	70	60	45	40	35
10.0–20.0	80	80	80	45	40
>20.0	100	100	100	100	100

Determine the permeability coefficient (the slope of the fitting straight line) according to the V-J (velocity-hydraulic gradient curve) of the sample under each working condition. The results are shown in Table 3.

Table 3. Permeability coefficient under different d_{10}

Constant	$C_u = 28.5$ $C_c = 0.64$				
Effective particle size (mm)	0.175	0.275	0.375	0.5625	0.75
Permeability coefficient *K* (cm/s)	0.0056	0.0248	0.0591	0.2653	0.8288

Then, we use different functions fit the permeability coefficient and d10, and we find that when the power function is used to fit the two variables, the correlation between the two variables is the highest, reaching 0.9963. Therefore, the fitting equation of K1 and d10 is:

$$K_1 = 1.957 d_{10}^{3.4044}$$

(2) Effect of C_u on permeability coefficient (K_2)

Using the same method as above, taking d10 = 0.375 and Cc = 0.83, the fitting equation of K2 and Cu is obtained as:

$$K_2 = -0.008 \ln C_u + 0.0641$$

(3) Effect of C_c on permeability coefficient (K_3)

Using the same method as above, taking $d_{10} = 0.375$ and $C_u = 20$, the fitting equation of K_3 and C_c is obtained as:

$$K_3 = 0.0023 C_C^2 - 0.007 C_c + 0.0336$$

(4) Formula establishment of Sandy pebble soil permeability coefficient (K)

From the above research, we can see that K has a nonlinear relationship with d_{10}, C_u, C_c. Based on the above conclusions, an empirical formula model is established as:

$$K = aK_1K_2K_3 + bK_1K_2 + cK_1K_3 + dK_2K_3 + eK_1 + fK_2 + gK_3 + h$$

Where a–h = model parameters.

Substituting 15 sets of data into the above formula, a linear equation system with K and 8 parameters is obtained. The equation system contains 15 linear equations. Solving the equation system through Matlab, the result is as shown.

$$K = -6.39 \times 10^4 K_1K_2K_3 - 411K_1K_2 + 377K_1K_3 - 74K_2K_3 + 77K_1 + 162K_2 + 154K_3 - 11$$

2.4 Sandy Pebble Triaxial Compressive Test

Through the triaxial compressive test (No drainage, axial load, strain control, strain rate is about 0.5%/min, confining pressure is selected as 100/200/400 MPa), the relationship between deviatoric stress and axial strain under different confining pressures is obtained, as shown in Fig. 3. And further draw the mohr's stress circle, as shown in Fig. 4.

According to the indoor triaxial compressive test, the macro-physical and mechanical parameters shown in Table 4 are obtained, which provides a reference for the calibration of numerical simulation parameters in the subsequent chapters.

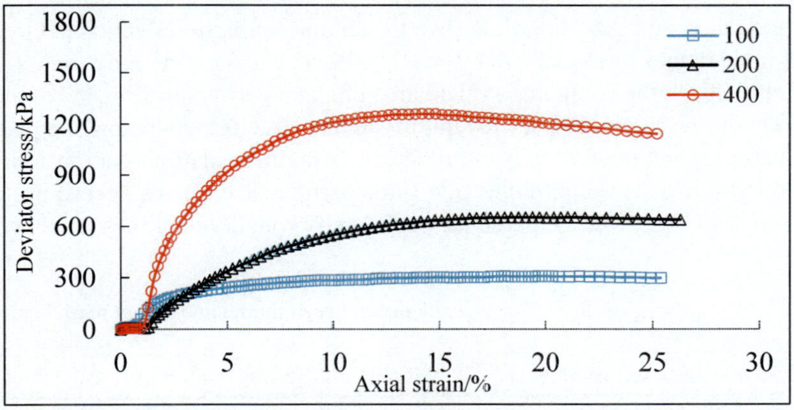

Fig. 3. Sandy pebble soil stress-strain curve under different confining pressure

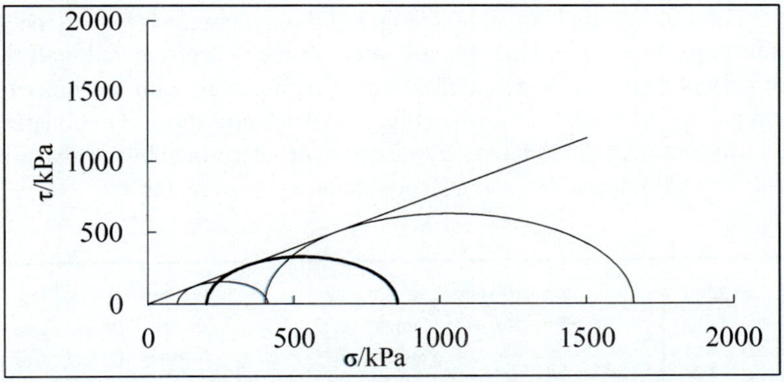

Fig. 4. Sandy pebble soil mohr's stress circle

Table 4. Macroscopic physical and mechanical parameters of Sandy pebble soil

Material name	Density (Kg/m^3)	Elastic Modulus (MPa)	Poisson's ratio	Internal friction angle	Cohesion (KPa)
Sandy pebble soil	2150	45	0.2	37.7	1.4

3 Numerical Modeling

3.1 Model Parameter Calibration

Use the PFC^{3D} software FISH language programming to build the model. The three-axis model should maintain the same meso-parameters as the subsequent engineering full-scale model, otherwise the parameter calibration through the triaxial compressive test will be meaningless. However, if the model is modeled according to the actual gradation and actual particle size, the amount of particles generated by the subsequent shield tunneling model will reach 100 million, and the software cannot calculate it. It can only match the computer's calculation efficiency by controlling the number of particles. Therefore, it is necessary to simplify and enlarge the gradation and particle size of the particles, but in order to better reflect the stratum gradation relationship, this article will roughly divide the particles into three particle size ranges according to the gradation curve in Fig. 2. And enlarge the particle size roughly ten times, as shown in Table 5.

Table 5. Correspondence between model and actual particle size

Actual particle size range	<20 mm	20–40 mm	>40 mm
Model particle size range	100–200 mm	200–400 mm	400–600 mm

Through the analysis of macro and micro parameters of sandy pebble soil, the simulation parameters suitable for the micro model were selected, and the corresponding numerical triaxial test was carried out by comparing with the indoor triaxial test (the comparison of results is shown in Fig. 5) to determine the simulation parameters. In addition, the permeability related parameters were determined by the analysis of permeability test. The parameters of each micromodel are shown in Table 6.

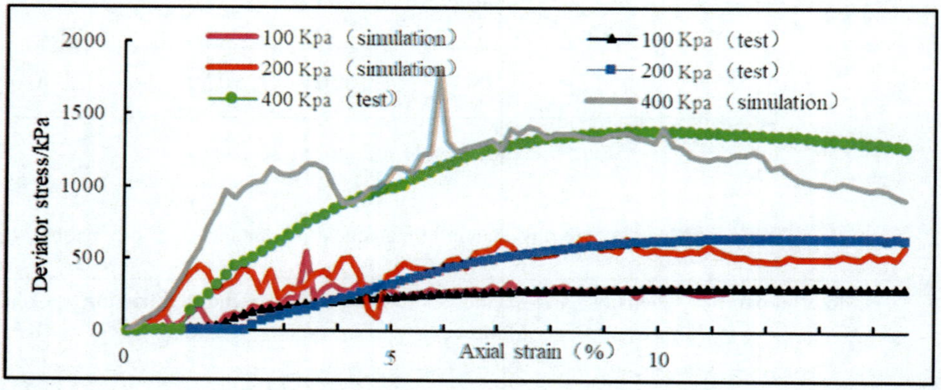

Fig. 5. Deviatoric stress-axial strain relationship diagram

Table 6. Sandy pebble soil Microscopic physical and mechanical parameters

Parameter name	P Kg/m^3	R/m	μ	K_n/(N/m)	K_t/(N/m)	bs/Mpa	r	Pw Kg/m^3	K m/s
Value	3300	0.05–0.3	10	200	3	0.2	0.15	1000	0.016

3.2 Analysis Model Establishment

According to the range of 35 m (transverse) × 20 m (longitudinal) × 40 m (high) in the actual formation, and the water level is 3 m below the surface. Use the built-in CFD of PFC and Python script to set the flow field of the corresponding water level, the flow pressure gradient is 0.02 m, the analysis model as shown in Fig. 6 is established.

Combining the actual engineering and simulation conditions, through the control variable method, a total of 13 full-scale simulation conditions are set for different factors, as shown in Table 7.

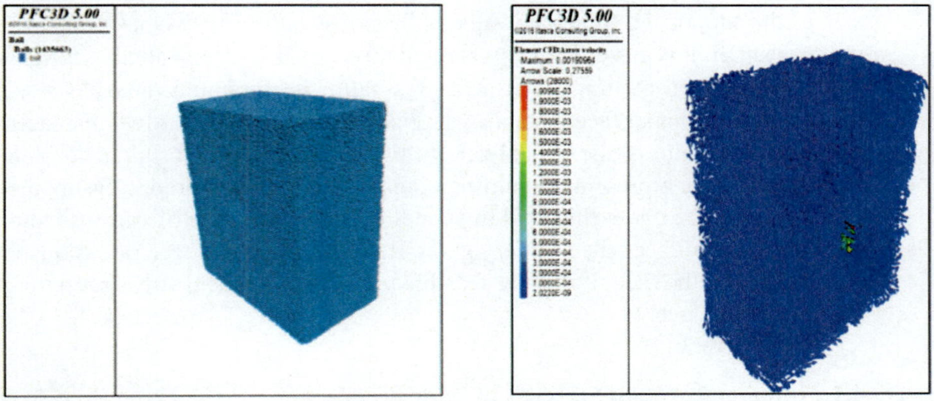

(a) Model particle generation diagram (b) Seepage flow field setting diagram

Fig. 6. Analysis model

Table 7. Full-scale simulation conditions

Serial number	Influencing factors					
	Stress ratio	D (m)	C (m)	C/D	Boulder	Remarks
1	0.5	6	15	2.5	None	Basic conditions
2	0.1	6	15	2.5	None	Influence of stress ratio (λ)
3	0.2	6	15	2.5	None	
4	0.3	6	15	2.5	None	
5	0.4	6	15	2.5	None	
6	0.6	6	15	2.5	None	
7	0.7	6	15	2.5	None	
8	0.8	6	15	2.5	None	
9	0.5	9	15	1.7	None	Influence of shield diameter
10	0.5	15	15	1	None	
11	0.5	6	6	1	None	Influence of buried depth
12	0.5	6	30	5	None	
13	0.5	6	15	2.5	Exist	Influence of Boulder

4 Discussion and Analysis

The stability of the tunnel face is actually attributed to the relationship between the shield cutter head support pressure and the limit value of the tunnel face instability. In other words, when the support pressure is greater than the limit support pressure, the stability of the tunnel face can be guaranteed. Therefore, the subsequent analysis can study the influence of various factors on the limit support pressure according to this idea. Two concepts are introduced here: (1) Stress ratio (λ): The ratio of the cutter head support pressure to the original horizontal stress of the formation at the center of the cutter head (including water pressure when containing water). Compared with simply analyzing the size of the supporting force, the value of the stress ratio is between 0 and 1, the change range is small, it is easier to grasp its regularity, and it is more valuable for engineering application and promotion. (2) Critical stress ratio: the minimum value of stress ratio that does not cause tunnel face instability (significant increase in particle displacement and velocity). The value of the critical stress ratio multiplied by the original horizontal stress of the tunnel face represents the ultimate support pressure required to ensure the stability of the tunnel face under this working condition. The influence of various factors on the critical stress ratio can also be very good. Reflecting the influence of various factors on the stability of the face, it is more feasible to study the critical stress ratio to reflect the stability of the face.

4.1 Influence of λ on Stability of Tunnel Face

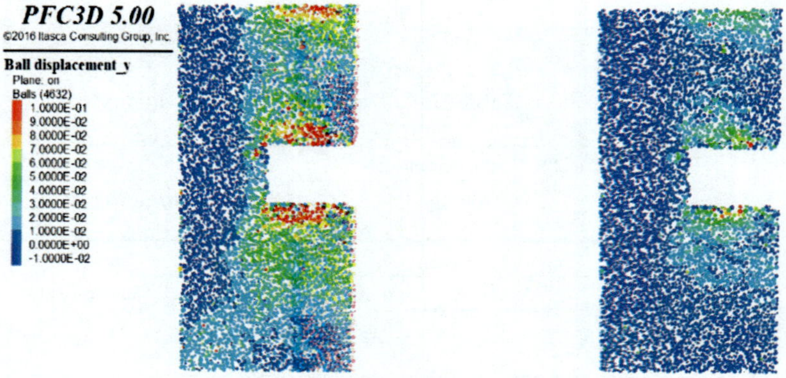

(a) λ=0.2 (b) λ=0.3

Fig. 7. Horizontal displacement of particles at different λ

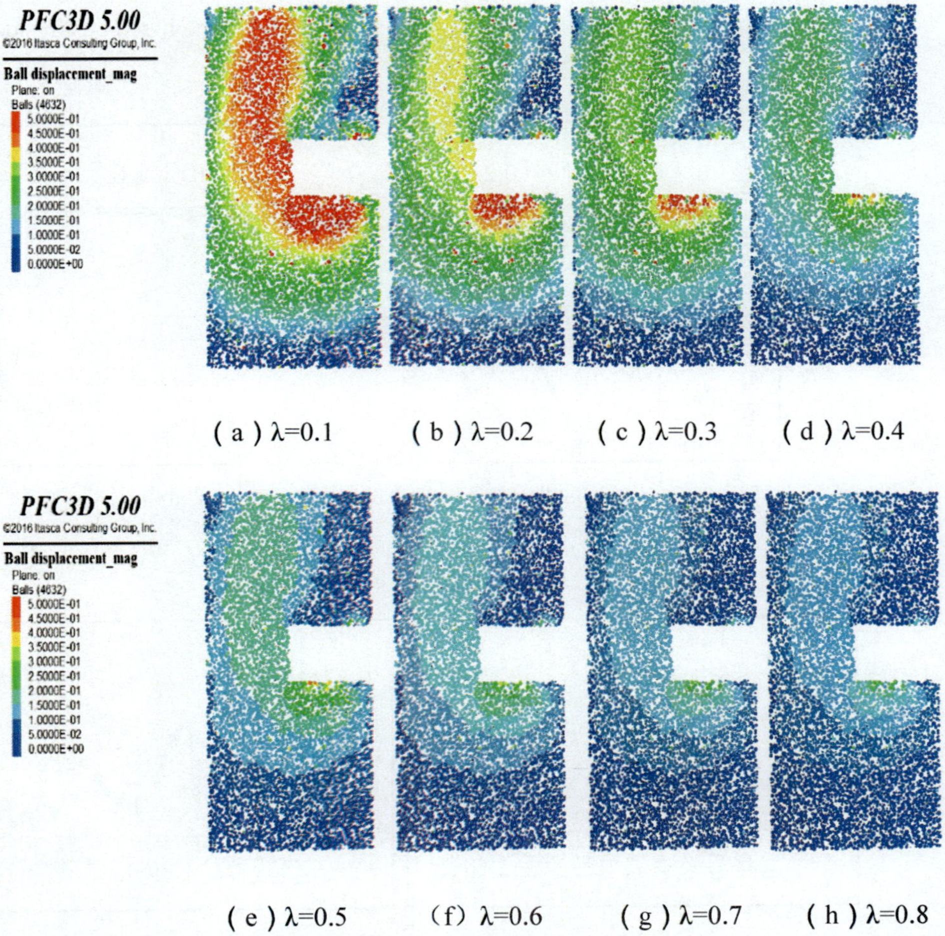

Fig. 8. Total displacement of particles at different λ

From Fig. 7, 8, we can get:

① When the stress ratio is between 0.2 and 0.3, the displacement of the particles changes greatly with the change of the stress ratio. When the stress ratio is greater than 0.3, the change of the displacement with the increase of the stress ratio is not obvious.

② There is a large displacement under the shield tunneling, because the flow field is considered and all the fluid grid outflow direction is the excavation surface. Therefore, during the excavation process, the particles below are constantly moving to the excavation surface under the action of the fluid drag force, resulting in a large displacement.

The particle displacement at the tunnel face is sorted, and the particle displacement values under different stress ratios are obtained as shown in Fig. 9. It can be seen from the figure that as the stress ratio increases, the particle displacement decreases gradually. When the stress ratio is greater than 0.3, the particle displacement value and the stress ratio basically show a linear relationship; when the stress ratio is 0.28, the curve appears an inflection point, and then "Turn down sharply", at this time it can be considered that the tunnel face is unstable.

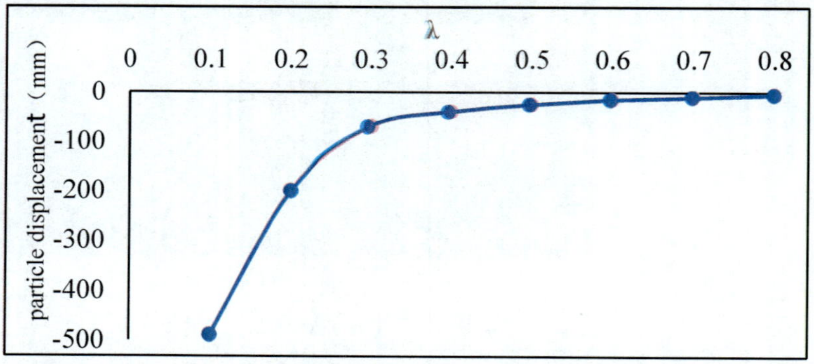

Fig. 9. Relationship between particle displacement and λ at the tunnel face

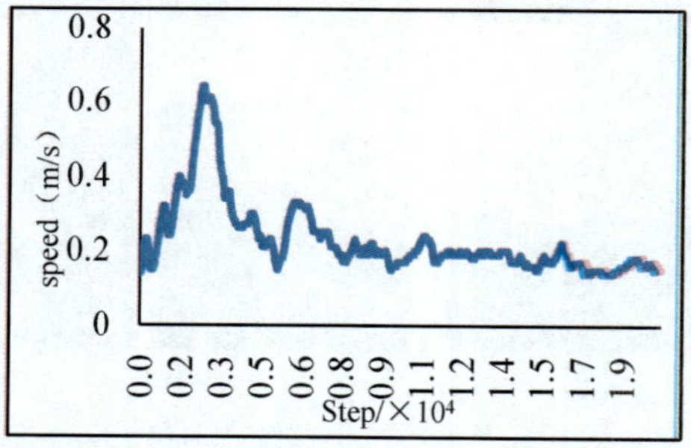

（a）λ=0.1

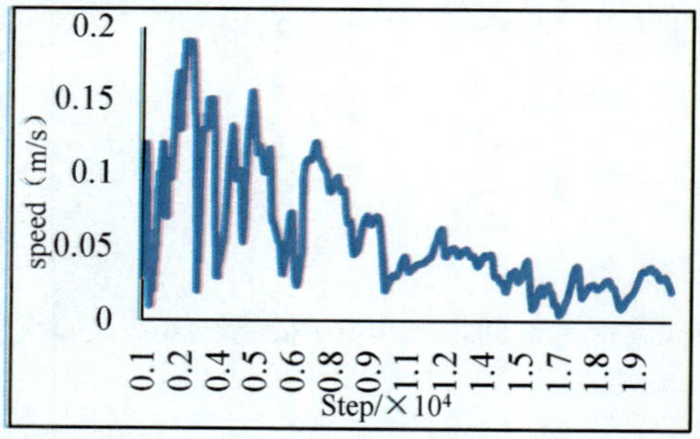

（b）λ=0.3

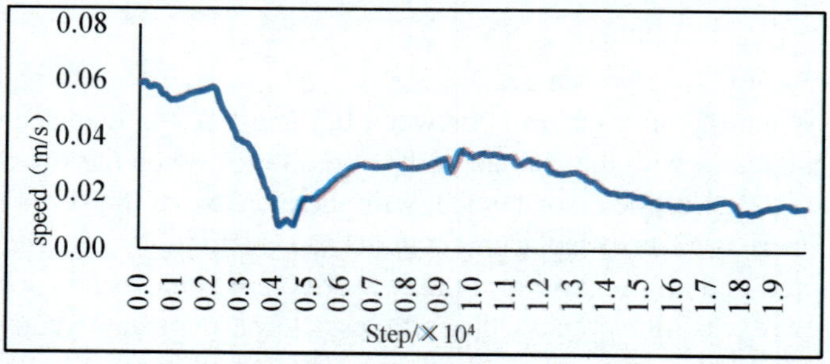

（c）λ=0.6

Fig. 10. Monitoring graph of particle moving speed at different λ

In addition to the particle displacement, the particle moving speed during the excavation process was also monitored. It was found that the change rule of the particle moving speed value is basically the same as the displacement change rule. The inflection point also occurs near the stress ratio 0.28, as shown in Fig. 10. Therefore, the critical stress ratio for the working condition of 15 m buried depth and 6 m shield diameter can be set as 0.28.

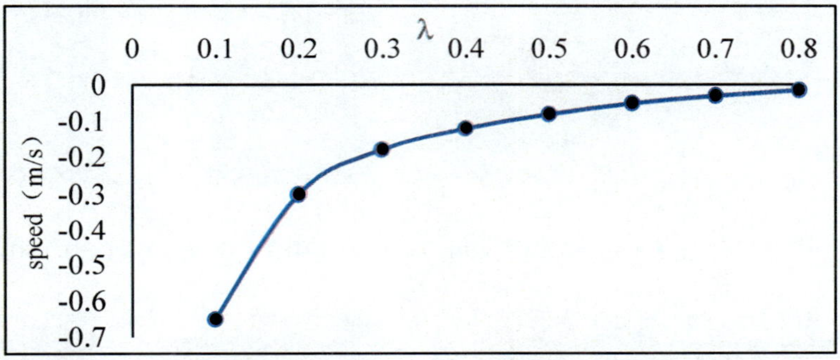

Fig. 11. Relationship between particle moving speed and λ

4.2 Influence of Shield Diameter on Stability of Tunnel Face

The section analyzes the critical stress ratio under different shield diameters (6 m, 9 m, 15 m, control buried depth is 15 m) to visually express the degree of influence. For the selection of the stress ratio in diameter 9 m and 15 m, set it with reference to diameter 6 m (working condition 1–8). The specific results are shown in Fig. 12, 13, 14, 15.

From the comparison of Fig. 12, 13, 14 and Fig. 7, 8, 9, 10, it can be seen that the larger the shield diameter, the larger the particle displacement and velocity under the same stress ratio. This shows that as the shield diameter increases, under the condition of constant burial depth, the self-stability ability of the face is weakened, and the risk of instability of the face increases.

Using the same method as Fig. 9 and 11, the critical stress ratios under the two working conditions of diameter 9 m and 15 m were obtained, and the relationship between the critical stress ratio and the shield diameter was drawn, as shown in Fig. 15.

It can be seen from Fig. 15 that the critical stress ratio also increases approximately linearly with the increase of the shield diameter, and because the buried depth is the same, the ultimate support pressure also increases linearly with the shield diameter. Therefore, when selecting shield diameter and tunneling parameters, the ultimate support pressure should be considered to avoid excessive shield diameter and mismatch of support pressure, which may cause instability of the tunnel face.

4.3 Influence on Buried Depth on Stability of Tunnel Face

This section analyzes the critical stress ratio under different buried depths (6 m, 15 m, 30 m, control shield diameter is 6 m) to visually present the effect of buried depth on the critical stress ratio and the stability of the tunnel face. The selection of related stress

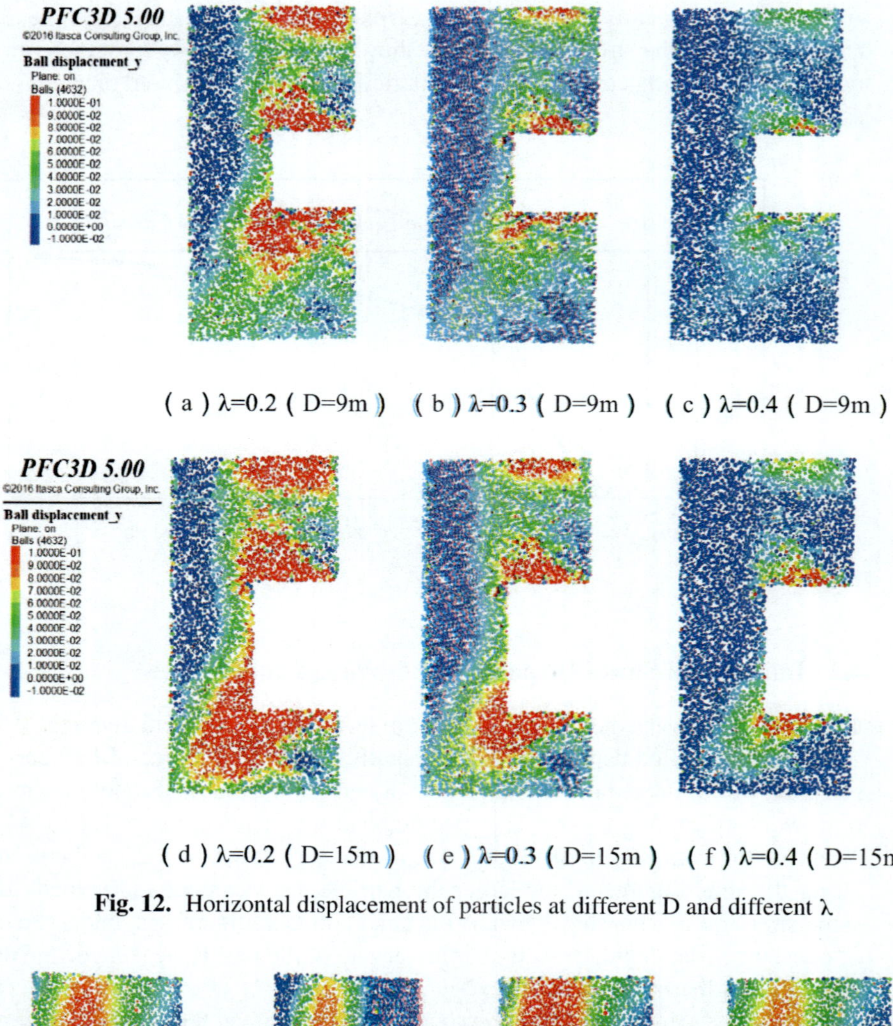

(a) λ=0.2 (D=9m) (b) λ=0.3 (D=9m) (c) λ=0.4 (D=9m)

(d) λ=0.2 (D=15m) (e) λ=0.3 (D=15m) (f) λ=0.4 (D=15m)

Fig. 12. Horizontal displacement of particles at different D and different λ

(a) λ=0.2 (D=9m) (b) λ=0.3 (D=9m) (c) λ=0.2 (D=15m) (d) λ=0.3 (D=15m)

Fig. 13. Total displacement of particles at different D and different λ

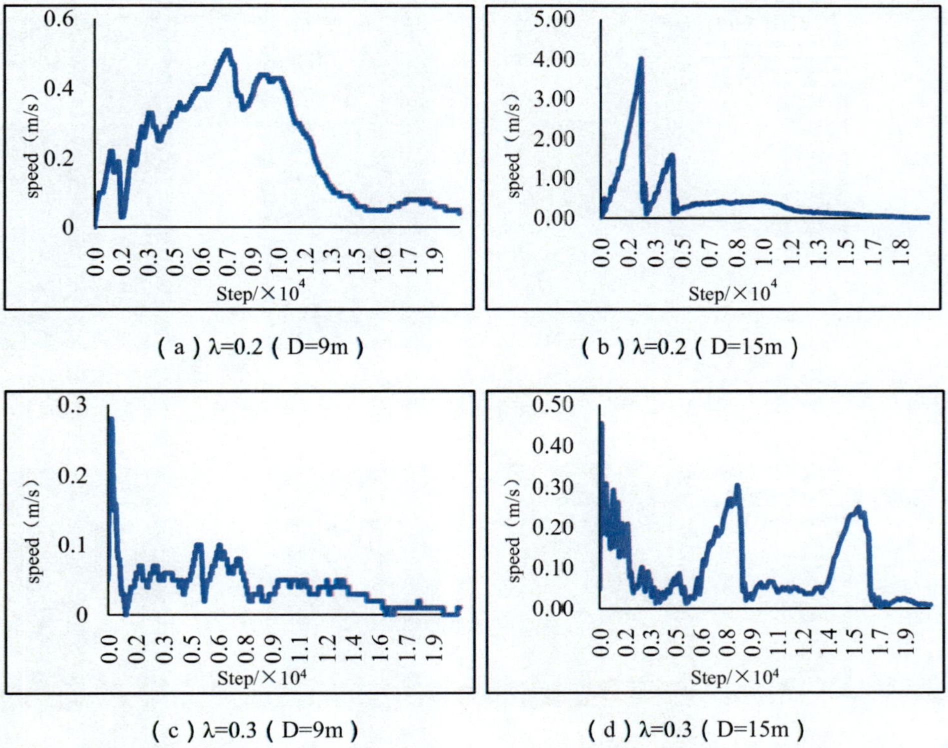

Fig. 14. Monitoring graph of particle moving speed at different D and different λ

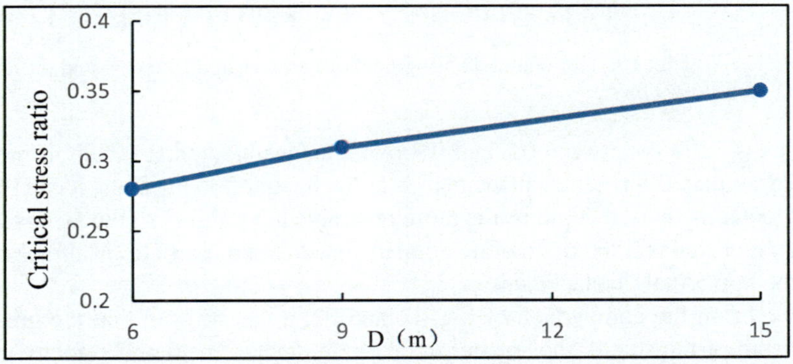

Fig. 15. Relationship between critical stress ratio and D

ratio still refers to working conditions 1–8. The specific results are shown in Fig. 16, 17, 18.

We can find that under the same stress ratio, the larger the buried depth is, the smaller the horizontal displacement of particles is, from Fig. 16 and 7. And we also find when displacement changes suddenly at the depth of 6 m, the stress radio is between 0.3 and 0.4. Moreover, when displacement change suddenly at the depth of 15 m and 30 m, the

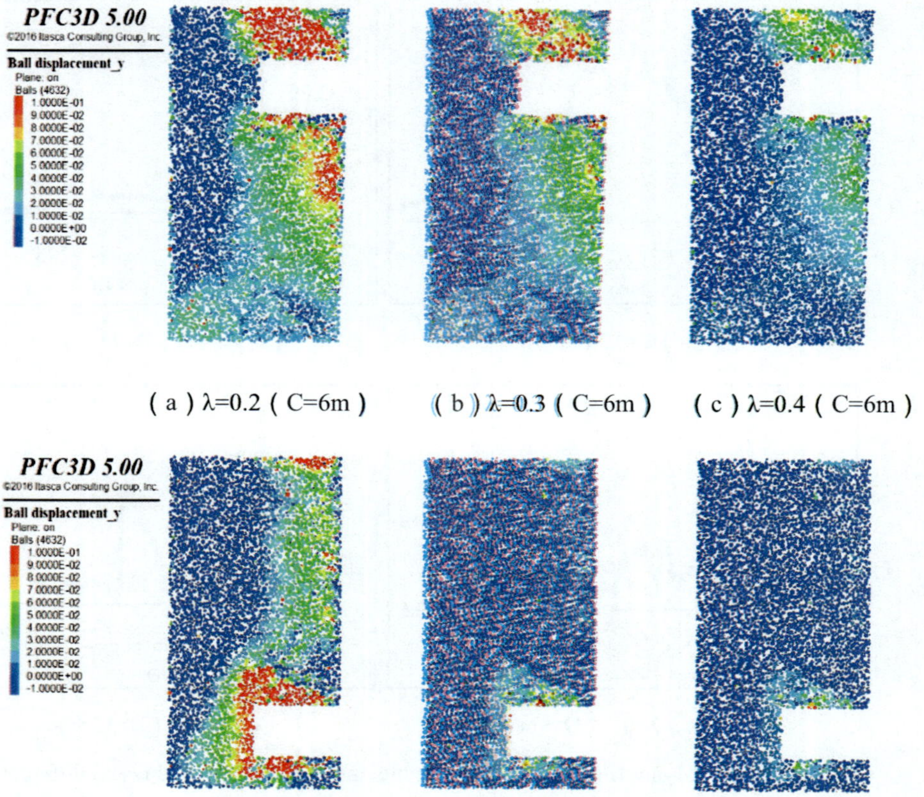

(d) λ=0.2 (C=30m) (e) λ=0.3 (C=30m) (f) λ=0.4 (C=30m)

Fig. 16. Horizontal displacement of particles at different C and different λ

stress radio is between 0.2 and 0.3. And the change gonna be tiny if the stress ratio is larger than 0.3 when displacement is at the depth of 30 m. Thus, it can referred that the greater the buried depth is, the more favorable the stability of the face is. And it will not be repeated that the trend of the total displacement of particles is almost consistent with the horizontal displacement.

From the comparison of Fig. 17 and 10, it can be seen that the greater the buried depth of the shield, the smaller the particle displacement and velocity under the same stress ratio. This indicates that as the buried depth increases, the relative stability of the tunnel face increases.

This section uses the same method as Fig. 9 and 11 to obtain the critical stress ratio of different buried depths, and draws the relationship between it and the buried depth, as shown in Fig. 18.

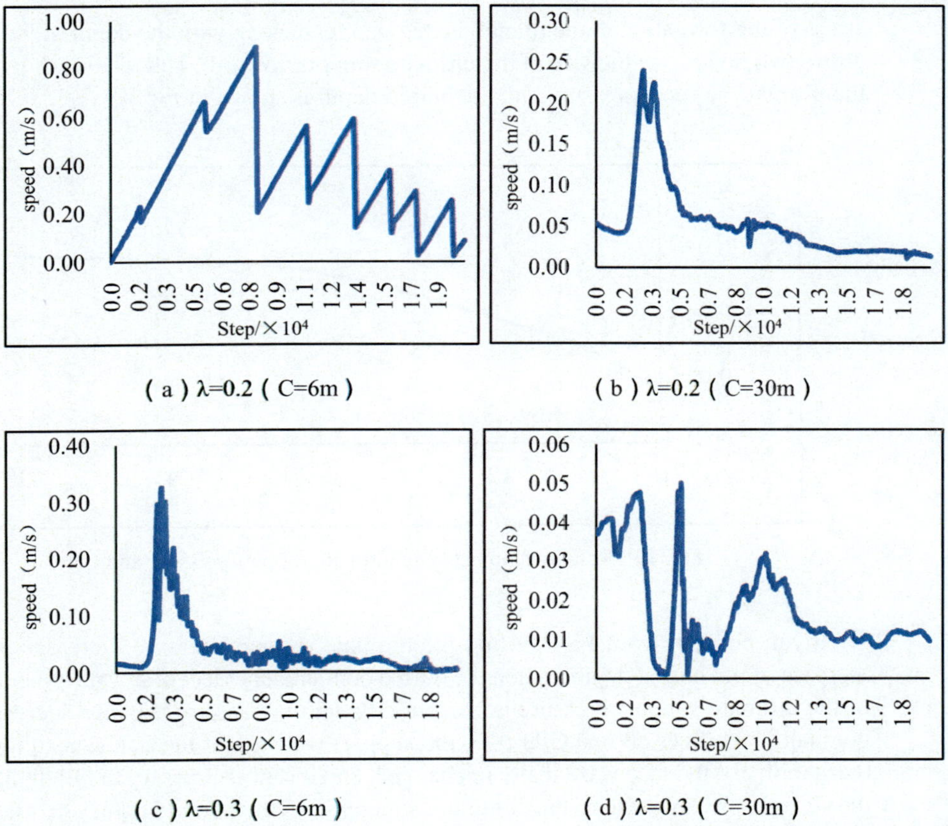

Fig. 17. Monitoring graph of particle moving speed at different C and different λ

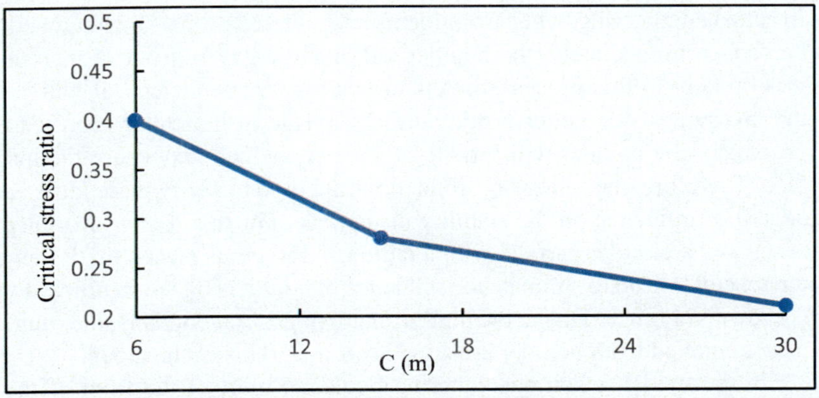

Fig. 18. Relationship between critical stress ratio and C

It can be seen from Fig. 18 that as the buried depth increases, the critical stress ratio gradually decreases, but the decreasing trend slows down and basically stabilizes at about 0.2. Although the critical stress ratio is gradually decreasing, the horizontal stress of the formation at the tunnel face is also increasing with the depth of burial. The above two determine the size of the ultimate support pressure. The relationship between the ultimate support pressure and the buried depth is shown in Fig. 19.

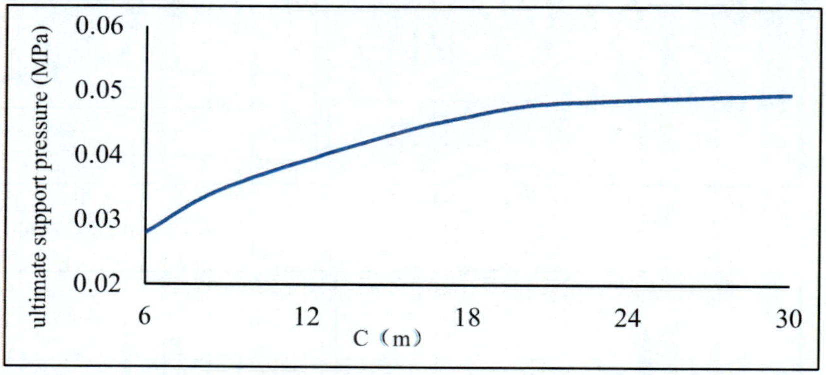

Fig. 19. Relationship between ultimate support pressure and C

It can be seen from Fig. 19 that the ultimate support pressure increases with the increase of the buried depth. When the buried depth reaches a certain value, the increase of the ultimate support pressure as the buried depth increases gradually decreases and gradually stabilizes. When $C/D > 3$, the support pressure of the cutter-head to ensure the stability of the face is basically unchanged, and it is not necessary to blindly increase the support pressure due to the continuous increase of the buried depth.

4.4 Influence of Boulders on Stability of Tunnel Face

In actual engineering, when a boulder is encountered during shield tunneling and cannot be broken immediately, the boulder will always stay in front of the cutter head until it is broken. In this process, the cutter head is "accumulated". It will greatly increase the pressure of the cutter head, causing damage to the cutter head, and the crushing process of the boulder will form a large cavity, which may cause the instability of the face. Therefore, the following simulations are used to verify these ideas and analyze the degree of influence on the stability of the face. For this reason, working condition 13 is set, and a circular particle with a radius of 150 cm is placed on the center line of the model cutter head to simulate the influence of boulders on the stability of the tunnel face is shown in Fig. 20. Due to the high strength of general boulders, the limit of the contact force removed from boulder crushing is set at 250 kN in the model.

In the model, when encountering boulders, most of the front of the face will be occupied by boulders, but when the contact force is not enough to break it instantly, the boulders will always stay in front of the cutter head until the set contact force limit is reached. The cutter head and the boulder are in point contact. To achieve the crushing condition, the cutter head pressure will be greatly increased, as shown in Fig. 21.

It can be seen from Fig. 21 that when the excavation reaches about 9 m, the cutter head starts to contact the boulder. At this time, the cutter head pressure rises rapidly.

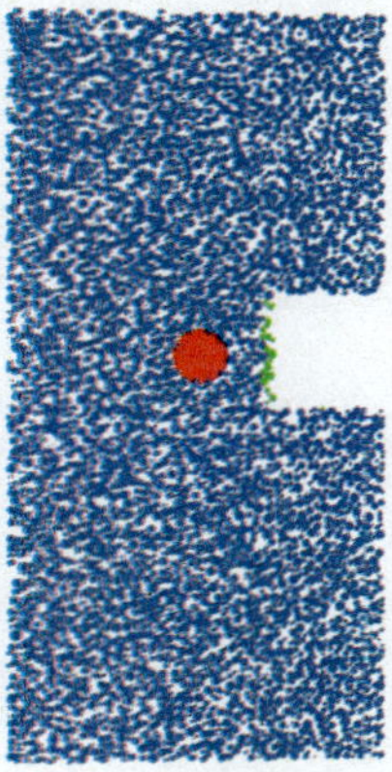

Fig. 20. Schematic diagram of boulder location

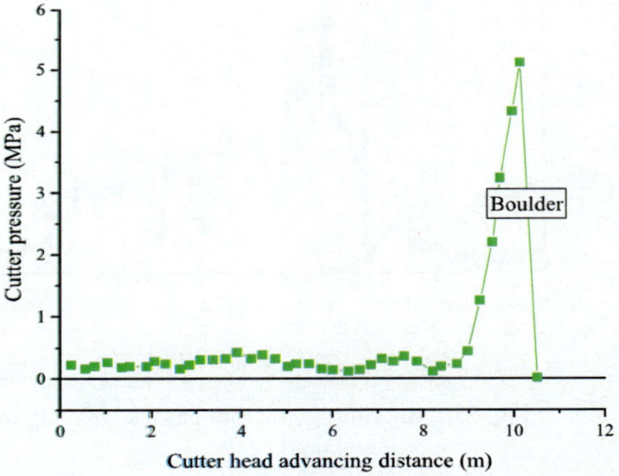

Fig. 21. Change of cutter head pressure during tunneling

After the pressure reaches the peak, it immediately decays to a value close to 0. This is because the boulder instantly It was deleted because there was no particle filling at its original location, and the contact between the cutter head and the particles decreased rapidly. This result verified the previous guess.

Comparing Fig. 22 with Fig. 8-e, it is found that the presence of boulders greatly increases the displacement of particles on the face of the tunnel, and has a greater impact on the particles of the entire formation, which greatly increases the risk of instability of the face.

It can be seen from Fig. 23 that the instantaneous horizontal velocity of the central particle of the tunnel face after the boulder is deleted is close to 3000 mm/s. Combined with Fig. 21, it can be judged that the tunnel face is at risk of instability. In general, the presence of boulders in front of the master is extremely dangerous for shield tunneling, and additional attention should be paid during construction.

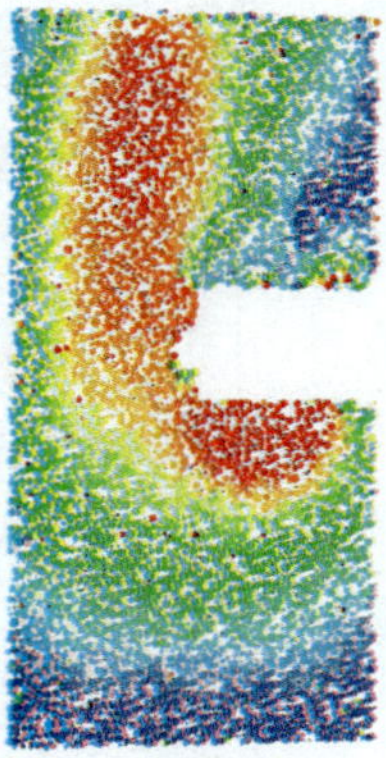

Fig. 22. Stratum displacement map after the boulder is broken

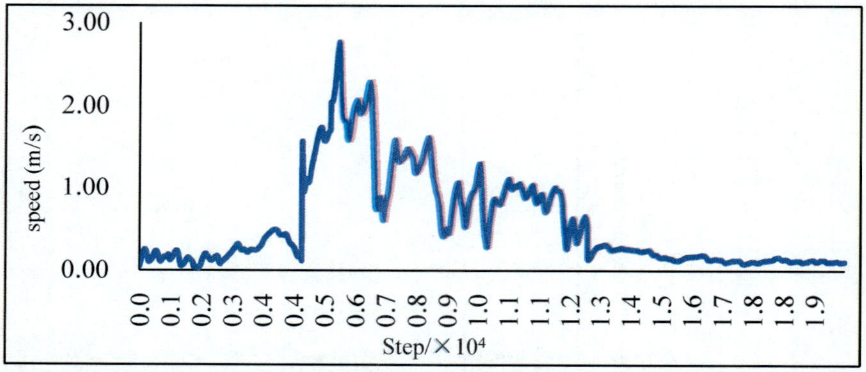

Fig. 23. Monitoring graph of particle velocity in the face

However, it needs to be pointed out that the relevant analysis in this section takes into account the engineering characteristics of sandy pebble, particle amplification effect and water seepage. Therefore, compared with the previous research results (basically between 0.15 and 0.3) obtained by most scholars in sand and other strata, the critical value obtained in this paper is higher, fluctuating around 0.3.

5 Conclusion

In this paper, combined with indoor experiments, using PFC3D to establish a numerical model, analyze the main factors affecting the stability of the shield excavation face in sand and gravel formations and their changing laws. The main conclusions obtained are as follows:

(1) The physical and mechanical properties of 5 sand pebble soil samples were tested in laboratory, and the empirical formula of permeability coefficient of sand pebble soil was obtained. PFC^{3D} mesoscopic model parameters, such as particle stiffness, density, friction coefficient and permeability coefficient, were calibrated according to laboratory test results.

(2) An indicator that can reflect the stability of the face is introduced: stress ratio (support pressure/horizontal stress of the original formation), and the influence of factors such as shield diameter, buried depth, and boulders on it is analyzed. The larger the stress ratio, the smaller the particle displacement and particle moving speed, and the more stable the formation during shield tunneling. When the stress ratio is less than the critical stress ratio, the tunnel face will lose stability.

(3) The critical stress ratio decreases with the increase of C/D, and when C/D reaches 3, the ultimate support pressure basically stabilizes.

(4) The presence of boulders greatly increases the risk of instability of the face, and the center of the face is broken immediately Particle speed can reach 3 m/s.

This paper considers the seepage problem relatively simply, such as the rough setting of the pressure gradient and the single direction of fluid inflow. On the basis of this paper, subsequent research work will consider the impact of water level changes on soil stability to study the impact of complex flow fields on shields, so as to obtain flow field parameters that are more in line with actual conditions, which will have greater guiding significance for the project.

References

1. Vermeer, P.A., Ruse, N.M., Marcher, T.: Tunnel heading stability in drained ground. Felsbau **20**(6), 8–18 (2002)
2. Zhang, C.P., Han, K.H., Zhang, D.L.: Face stability analysis of shallow circular tunnels in cohesive–frictional soils. Tunneling Underground Space Technol. **50**(8), 345–357 (2015)
3. Melis Maynar, M.J., Medina Rodriguez, L.E.: Discrete numerical model for analysis of earth pressure balance tunnel excavation. J. Geotech. Geoenviron. Eng. **131**(10), 1234–1242 (2005)
4. Chen, R.P., Tang, L.J., Ling, D.S., et al.: Face stability analysis of shallow shield tunnels in dry sandy ground using the discrete element method. Comput. Geotech. **38**(2), 187–195 (2011)
5. Karim Asm, M.: Three-dimensional discrete element modeling of tunneling. University of Alberta, Alberta (2007)
6. Wang, M.M., Wei, L.H., Lu, J.F.: Study of face stability of cobble-soil shield tunneling at Chengdu metro. Rock Soil Mech. **31**(1), 99–105 (2011)
7. Kirsch, A.: Experimental investigation of the face stability of shallow tunnels in sand. Acta Geotech. **5**(1), 43–62 (2010)
8. Chen, R.P., Li, J., Chen, Y.M.: Large-scale tests on face stability of shield tunneling in dry cohesionless soil. Chin. J. Geotechnical Eng. **33**(1), 117–122 (2011)
9. Lv, X.L., Zhou, Y.C., Li, F.D.: Centrifuge model test and numerical simulation of stability of excavation face of shield tunnel in silty sand. Rock Soil Mech. **37**(11), 3324–3328 (2016)
10. Sun, X.H., Liao, L.C., Lin, H.S.: Numerical simulation research on excavation face stability of different depths of shield tunnel. J. Southeast Univ. **47**(1), 164–169 (2017)
11. Lin, H.S.: Study on the face stability around shield tunnels and numerical simulation of PFC. Southeast University, Jiangsu (2016)
12. Li, R.J.: Stability analysis of soil pressure balance shield tunneling face in sand stratum. Beijing Jiaotong University, Beijing (2019)
13. Liu, G.L.: Research on Geological Property and Deformation Mechanism of Sandy Cobble Stratum in Chengdu Metro. Southwest Jiaotong University, Sichuan (2016)

The Influence of Recycled Power Dust on Asphalt Mixture

Qingqing Zhang[1](✉), Lingyu Meng[1], and Jinyan Liu[2,3]

[1] Cangzhou Jingtong Engineering Testing Co. Ltd., Cangzhou 061000, China
[2] Cangzhou Municipal Engineering Company Limited, Cangzhou 061000, China
[3] Hebei Province Road Materials and Technology Engineering Technology Research Center, Cangzhou 061000, China

Abstract. Laboratory tests aimed at investigating compaction properties, volumetric characteristics, and mechanical performance of the bituminous mixtures. The results suggest that recycled powder dust is suitable to be used in bituminous mixtures, also, in certain conditions, the investigated recycled powder dust increases the performance of the corresponding mixtures in comparison to mineral filler. For AC-20C asphalt mixture, when the ratio of the recycled powder dust substitution is below 50%, the water stability and low temperature stability of the mixture meet the specification requirements. When the ratio of the substitution exceeds 75%, the mixture is found prone to rutting. It is recommended to use the recycled powder dust in medium and light traffic conditions.

1 Introduction

In the production process of an asphalt mixture, a large amount of dust is generally collected due to heating of and dust removal from aggregates. This dust is mainly a mixture of mineral powders and impurities. If this dust is recycled, it not only preserves a large amount of resources, but also reduces the environmental pollution. Recycling reduces the need for extracting, refining, and processing of raw materials. All of these create substantial air and water pollution. One of the cheapest and most effective ways to eliminate such waste and protect the environment is recycling in hot mix asphalt (HMA) mixture. Technical Specification for Highway Asphalt Pavement Construction (JTG F40-2004) [1] stipulates that the aggregate dust can be used as part of the mineral powder, but the amount shall not exceed 25% of the total amount of filler. In addition, the plasticity index of the dust shall not exceed 4%. Mo Luqun [2] documented that the performance of an asphalt mixture has no significant variation when the dust content is less than 50%. Filler is an integral component of asphalt mixtures, and its content greatly affects mechanical properties of such mixtures. If the filler content is too high, the bituminous mixture can be weakened considerably as the filler influences the optimum content of bitumen, and mixture constructability, and mixture's elastic properties.

To evaluate the impact of the recycled powder dust on asphalt mixtures under different mineral filler rate of substitution, this paper studied the following three aspects:

Y. Liu et al. (Eds.): GeoChina 2021, SUCI, pp. 132–141, 2021.
https://doi.org/10.1007/978-3-030-80316-2_12

(1) the difference between the recycled power dust and mineral filler in terms of particle size analysis and microscopic characteristics.
(2) the impact of the recycled powder dust on composite shear modulus (G*) and phase angle (δ).
(3) the low temperature stability, water stability, and the effect of high temperature and water on the asphalt mixture.

2 Material Properties of Recycled Power Dust and Mineral Filler

2.1 Conventional Performance Index Properties

The limestone mineral filler used in this experiment was produced in Hebei China. The recycled power dust was obtained directly from the dust bag at the production site. According to the test procedure [3], the performance index properties of the recycled powder dust and mineral filler were measured. The results are shown in Table 1.

Table 1. Performance index properties of recovered power dust and mineral filler

Performance	Recycled powder dust	Mineral filler	Urban expressway, highway, first-class highway standard [1]	Other grades of road standards [1]
Apparent density (t/m^3)	2.750	2.783	≥2.5	≥2.5
Hydrophilic coefficient	0.52	0.6	<1	<1
Liquid limit (%)	36.1	21.7	/	/
Plastic limit (%)	26.1	18.7	/	/
Plasticity index (%)	10	3	<4	<4
Methylene blue (g/kg)	6.0	4.5	<25	/
Heating stability	No change in color	No change in color	/	/

2.2 Scanning Electron Microscope Test

Using the electron microscope BCPCAS4800, the recycled power dust and mineral filler were scanned, and the microstructures are presented in Fig. 1, 2, 3 and 4.

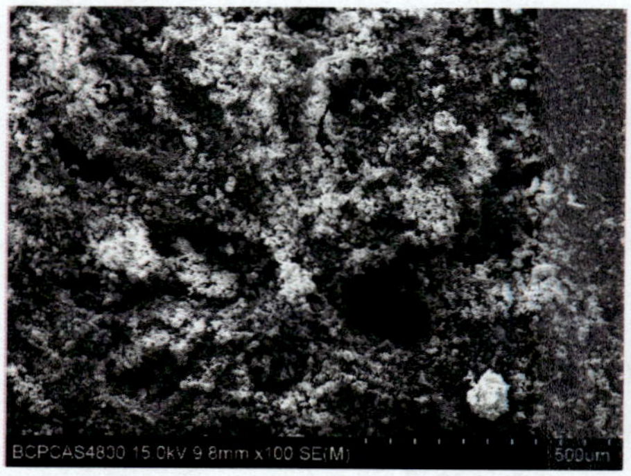

Fig. 1. Recycled power dust magnified 100 times

Fig. 2. Mineral filler magnified 100 times

Fig. 3. Recycled power dust magnified 300 times

Fig. 4. Mineral filler magnified 300 times.

2.3 Specific Surface Area and Particle Size Analyses

According to the "Cement Specific Surface Area Measurement Method Blaine Method" (GB/T 8074-2008) [4], the specific surface area of the recycled power dust and mineral filler was measured. The laser particle size analyzer was used to perform particle size analyses [5] based on the physical phenomenon. It is the most widely used technique for the particle size analysis and has successfully been used for powdered materials because it offers accurate, reproducible, and high resolution light scattering measurements to ensure full characterization of the particle size distribution.

When the beam is blocked by particles, a part of light scatters. The propagation direction of the scattered light forms an angle θ with the propagation direction of the main beam. The magnitude of the scattering angle θ is related to the size of the particles. The larger the particle, the smaller the θ angle of the scattered light. By measuring the intensity of the scattered light at different angles, the particle size distribution of the sample is obtained. Usually D50 is used to indicate the average particle size of a powder. D50 is the particle size corresponding to the cumulative particle size distribution percentage of a sample reaching 50%. The median is the diameter where half of the sizes are smaller than this value and half are larger. The median is called by various names including d50, dv50, or x50. The median is the most stable calculation generated by laser diffraction and shall be the value used for a single point specification in most cases. D90 represents the particle size when the cumulative particle size distribution percentage of a sample reaches 90%. Its physical meaning is that the particle size is smaller than 90%. The test results of particle size analysis are shown in Table 2.

Table 2. Particle size analysis of recycled powder dust and mineral filler

Performance	Recycled powder dust	Mineral filler
D50 (μm)	6.2	62
D90 (μm)	18	510
Specific surface area (m^2/kg)	671	453

2.4 Analyses

(1) The apparent density, hydrophilicity coefficient, methylene blue value, and heating stability of the recycled powder dust met the technical requirements of the specification. The plastic index (Table 1) is quite different from the specification. The apparent density and hydrophilic coefficient of the recycled powder dust are smaller than that of mineral filler.

(2) It can be seen from the electron microscope scanning results that the particle size of the recycled powder dust is significantly smaller than that of the mineral filler.

(3) From the results of the particle size analysis, the average particle size of the recycled powder dust is 6.2 μm, which is about 1/10 of the average particle size of the mineral filler. The specific surface area of the recycled powder dust is smaller than that of mineral filler.

3 Effects of Recycled Powder Dust and Mineral Filler on Asphalt Mixture

The composition structure of an asphalt mixture has a significant effect on the high temperature stability. In general, the contribution rate of asphalt binder to the rutting resistance of an asphalt mixture is up to 40% [6]. To study the influence of the mineral filler and the recycled powder dust on the high temperature performance of asphalt binder, asphalt mixtures were prepared using three different filler-to-bitumen ratios, i.e., 0.8, 1.0, and 1.36. The rheological properties of asphalt mortars were tested by the dynamic shear rheometer (DSR) method [7] and compared with Binzhou Zhonghai's No. 70 road asphalt. The ratios of the recycled powder dust in the prepared asphalt mortar to the mass of mineral powder was 0%, 50%, and 100%, where 0% means no recycled powder dust is added and all fillers use mineral powder.

3.1 Preparation of Asphalt Mortar

The amounts of powder and asphalt according to the filler-to-bitumen ratio of the asphalt mortar were calculated. The mineral filler, recycled powder dust, and No. 70 original asphalt were conditioned in an oven at 150 °C. When the powder is added to the asphalt at one time, because the amount of powder is large, it is easy to agglomerate under the viscosity of the asphalt. Thus, it is advisable to add the powder slowly. At the same time, high-speed shearing apparatus was used to mix the powder and asphalt for one hour at 150 °C and at rotation speed of 6000 rad/min to ensure that the powder is evenly dispersed in the asphalt.

3.2 DSR Test Results

The DSR test results of No. 70 road petroleum asphalt are shown in Table 3, and the DSR test results of asphalt mortar are shown in Figs. 5, 6, 7 and 8.

Table 3. DSR test results of No. 70 original asphalt

Sample	Temperature (°C)	Composite shear modulus G* (Pa)	Phase angle δ (°)	Rutting factor $G^*/\sin\delta$ (kPa)
No. 70 Original asphalt	46	18240	83.22	18.36
	52	7515	85.26	7.541
	58	3256	87.19	3.26
	64	1438	89.38	1.438
	70	683.2	92.11	0.6837

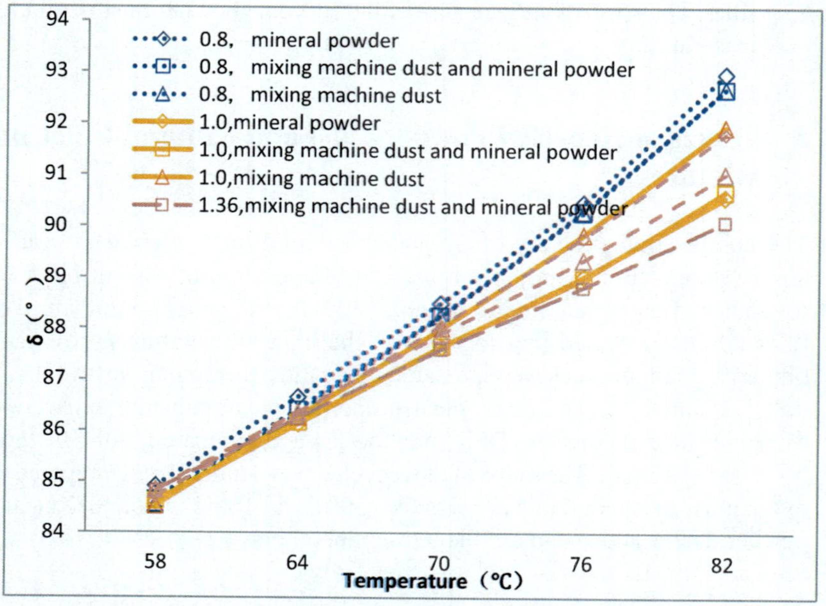

Fig. 5. Relationship between δ and temperature

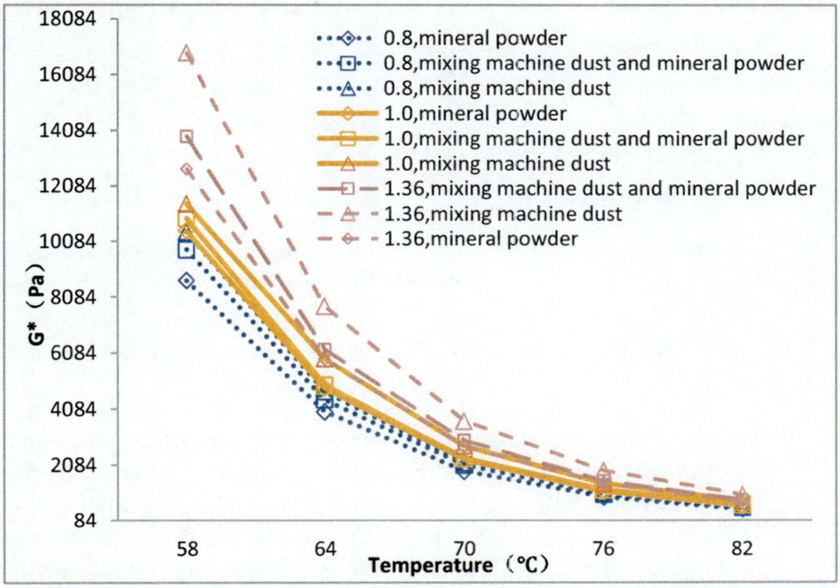

Fig. 6. Relationship between G* and temperature

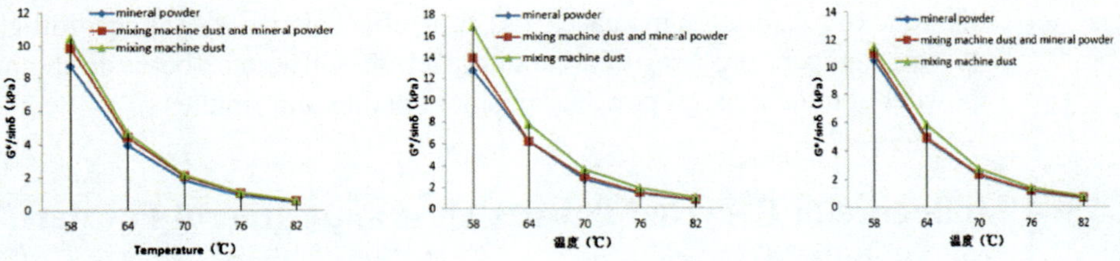

(a) Filler-to-bitumen ratio is 0.8 **(b) Filler-to-bitumen ratio is 1.0** **(c) Filler-to-bitumen ratio is 1.36**

Fig. 7. Relationship between G*/sinδ and temperature under different recycled powder dust content

3.3 Analyses

(1) As the temperature rises within a certain temperature range, the high-temperature stability of asphalt mortar and original asphalt have a gradual decline, but the composite shear modulus G* of asphalt mortar is higher than the virgin (original) asphalt. It indicates that the addition of mineral powder and recycled powder dust has improved the stiffness of asphalt binder. The degree of the increase corresponds to the filler-to-bitumen ratio, as the ratio of substitution increased, the composite shear modulus G* of the asphalt mortar increase. The degree of the increase for the recycled powder dust is much higher than the mineral powder filler.

(2) As the temperature increases, the phase angle δ of asphalt mortar shows an upward trend as compare to the virgin (original) asphalt. Under the same filler-to-bitumen

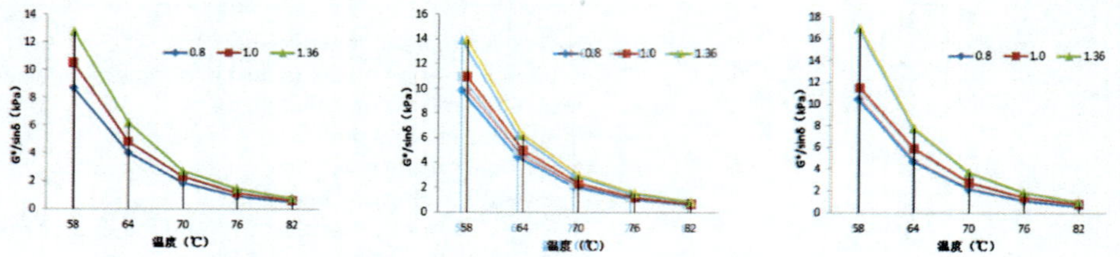

(a) 100%mineral powder

(b) 50%mineral powder+50% recycled powder dust

(c) 100% recycled powder dust

Fig. 8. Relationship between G*/sinδ and temperature under different filler-to-bitumen ratios

ratio, as the ratio of substitution increases, the value of the phase angle δ decreases accordingly. This shows that the addition of mineral powder and recycled powder dust was able to improve the elastic properties of asphalt binder.

(3) The rutting factor G*/sinδ of the asphalt mortar increases with the increasing filler-to-bitumen ratio, and is larger than the original asphalt. This shows that the addition of recycled powder dust and mineral powder can improve the high-temperature stability and anti-rutting ability of the asphalt binder, and the more the amount added, the better the improvement effect; with the same filler-to-bitumen ratio, as the ratio of substitution increases, its lifting effect also increases accordingly. When the test temperature rises to a certain level, the difference between the recovered powder and the mineral powder becomes smaller and smaller.

4 Influence of Recycled Powder Dust and Mineral Powder Filler on Asphalt Mixture

4.1 Benchmark Mix Ratio

Asphalt binder from Binzhou Zhonghai's A-level No. 70 road petroleum asphalt was used in this study. Coarse aggregates are made of 10–20 mm, 5–10 mm, 3–5 mm limestone gravel produced in Baoding Mancheng. The fine aggregate is 0–3 mm limestone machine-made sand produced in Mancheng. Limestone mineral powder filler was used in experiments. All of the material property indicators of the above raw materials met the requirements of the "Technical Specifications for Construction of Highway Asphalt Pavements" JTG F40 [1]. According to the Marshall test mix proportion design method for AC-20C, the optimum asphalt content was found to be 4.0%. Tables 4 and 5 show the blending ratio of mineral materials and the synthetic gradation of mineral materials, respectively. The standard Marshall specimens were prepared at the optimal asphalt content of 4%. The mixing temperature of the mixture was 160 °C and the forming temperature was 150 °C. The volumetric parameters and mechanical properties of AC-20C met the specifications and the results are shown in Table 6.

Table 4. Blending of AC-20C asphalt mixture

Mixture type	10–20 mm (%)	5–10 mm (%)	3–5 mm (%)	Machine-made sand (%)	Mineral powder (%)	Oil-stone ratio (%)
AC-20C	38	19	18	23	2	4

Table 5. AC-20C asphalt mixture gradation

The mass percentage (%) passing the following sieve holes (mm)												
Mixture gradation	26.5	19	16	13.2	9.5	4.75	2.36	1.18	0.6	0.3	0.15	0.075
Synthetic gradation	100	95.2	831	75.9	64.1	38.3	21.8	17.4	12.5	9.3	7.4	5.9
Gradation range [1]	100	90–100	78–92	62–80	50–72	26–56	16–44	12–33	8–24	5–17	4–13	3–7

Table 6. Volumetric parameters and mechanical properties of AC-20C asphalt mixture

Oil-stone ratio (%)	Gross volume relative density	Void ratio (%)	Mineral clearance rate (%)	Asphalt saturation (%)	Stability (kN)	Flow value (mm)
4.0	2.43	5.5	14.23	67.65	8.54	2.51
Specification requirements [1]	—	4–6	≥14	65–75	≥8	1.5–4.0

4.2 Performance Tests

The experimental activity involved several tests selected to evaluate the volumetric characteristics and the performance-related properties of the bituminous mixtures, useful for assessing the role of the recycled power dust in replacement of the traditional mineral filler. Five different specimens were prepared, and the results are presented in Table 7. In Table 7, No. 1, No. 2, No. 3, No. 4, and No. 5 represent the recycled powder dust contents of 0%, 25%, 50%, 75%, and 100%, respectively. The experiments were carried out separately including Marshall test, water immersion Marshall test, freeze-thaw splitting test, and low temperature bending test [6, 7]. The typical dimension of specimen is with height of 62 ± 2 mm and diameter of 150 mm. The hamburger rut tests were carried out at water bath temperature of 50 °C, and the movement speed of the steel wheel was about 52 times/min. When the steel wheel reaches 20,000 reciprocating movements, it would be discontinued. The test results are shown in Table 7.

Table 7. Performance index properties of asphalt mixture

Detection indicator	Mixture type number					
	No. 1	No. 2	No. 3	No. 4	No. 5	Technical index
Void ratio (%)	5.5	5.4	5.4	5.5	5.1	4–6
Stability (kN)	8.54	9.41	9.37	9.83	10.16	≥8
Flow value (mm)	2.51	2.13	2.38	2.17	1.96	1.5–4
Residual stability (%)	96.6	92.8	84.2	82.4	81.3	≥80
Residual strength ratio (%)	80.5	81.1	84.9	81.1	78.0	≥80
Maximum bending strain (με)	2236.15	2119.27	2017.46	1986.72	1913.47	≥2000
Hamburg rut depth (mm)	−5.72	−7.78	−8.63	−8.25	−9.89	/

The observations are given as follows:

(1) The stability increases with increasing amount of recycled powder dust. However, the flow value yielded opposite trend.
(2) As the amount of recycled powder dust increases, the Marshall residual stability of asphalt mixture gradually decreases. The residual stability of asphalt mixture still met the requirements of the specification. However, when all the recycled powder dust replaces the mineral powder, the residual strength was not able to meet the specification requirements.
(3) When the Ratio of substitution is 25%, the residual strength is the highest. When the recycled powder dust partially replaced the mineral powder, it has little effect on the water stability of the asphalt mixture.
(4) Under low temperature conditions, with increasing of recycled powder dust content, the low temperature crack resistance of asphalt mixture gradually decreases, which is consistent with the research of Liu Weimin [8].
(5) According to the data of Hamburg rutting, under the combined action of high temperature and water, the addition of recycled powder dust has a significant impact on the performance of the asphalt mixture. As the ratio of substitution increases, the rutting depth of the asphalt mixture increase as well.

5 Conclusions and Recommendations

The comparative laboratory investigation described in this paper focused on the effects of recycled powder dust as filler in bituminous mixtures. The analysis aimed at investigating compaction properties, volumetric characteristics, and mechanical performance of the bituminous mixtures. The conclusions and recommendations are given as follows:

(1) From the perspective of microscopic and particle size analysis, the average particle size of the recycled powder dust is significantly smaller than the average particle size of the mineral powder.
(2) When the filler-to-bitumen ratio is ≤ 1.36, the recycled powder dust and mineral powder can increase the high-temperature stiffness of the asphalt mortar, and the higher the filler-to-bitumen ratio, the stronger the improvement, but the same quality of recovery powder has a better lifting effect than mineral powder, and as the Ratio of substitution increases, the high temperature stiffness of asphalt mortar is greater. For asphalt mortar, the greater the amount of recycled powder dust, the better.
(3) For AC-20C asphalt mixture, when the ratio of substitution was below 50%, the water stability and low temperature stability of the mixture was able to meet the specification requirements. When the ratio of substitution exceeded 75%, the mixture was prone to rutting. It is recommended to use the recycled powder dust in medium and light traffic sections. Cautions should be made to the local rainfall and road drainage.

References

1. JTG F40-2004: Technical specification for highway asphalt pavement construction
2. Mo, L., Xu, M.: Discuss the influence of mixing with different content of mixing plant recycled powder dust on the performance of asphalt mixture. Eng. Technol.
3. JTG E42-2005: Highway Engineering Aggregate Test Regulations
4. GB/T 8074-2008: Method for Determination of Specific Surface Area of Cement - Brinell Method. China Standard Press (2008)
5. GB/T 19077-2016/ISO 13320:2009: Particle Size Distribution Laser Diffraction Method. China Standard Press (2016)
6. Yu, S.: Specification and test of performance graded asphalt binder
7. JTG E20-2011: Highway engineering asphalt and asphalt mixture test regulations
8. Liu, W., Lin, Q.: Research on utilization of asphalt mixture recycled powder dust. 49–51 (2008)

Author Index

Y. Liu et al. (Eds.): GeoChina 2021, SUCI, pp. 143–144, 2021.
https://doi.org/10.1007/978-3-030-80316-2

Z